SEA KING

Series Editor : Christopher Chant

Foulis

Haynes

ISBN 0 85429 377 9

A FOULIS Aircraft Book

First published 1983

Published by:
Haynes Publishing Group
Sparkford,
Yeovil,
Somerset BA22 7JJ

Distributed in North America by:
Haynes Publications Inc.
861 Lawrence Drive,
Newbury Park,
California 91320, USA

Produced by:
Winchmore Publishing Services Limited,
40 Triton Square,
London, NW1 3HG

Picture research Jonathan Moore
Printed in Hong Kong by Lee Fung Asco Limited.

Further titles in this series will be published at
regular intervals. For information on new titles
please contact your bookseller or write to the
publisher

Contents

Though by no means a world-beater in terms of performance or payload, the Sikorsky S-61 series may justly be regarded as the Western world's most important helicopter. Evolved as the HSS-2 anti-submarine helicopter for the US Navy, the S-61 was soon in large-scale production as the SH-3 Sea King and under development into a wide variety of other roles. A sturdy airframe, turbine power, viceless handling characteristics and a relatively capacious fuselage all contributed to its success as a multi-role type in fields as far apart as logging, VIP transport and armed aircrew rescue. And there can be no surer indication of overall utility than worldwide demand, which the Sea King has certainly enjoyed. Extensive licence manufacture has been undertaken by Westland in the UK, Agusta in Italy and Mitsubishi in Japan.

Genesis

Among the many technical and tactical lessons to emerge from World War II was that of the threat posed to ships, both merchantmen and naval vessels, by the submarine. The threat to Allied sea lanes had been countered in World War II, but it had been a close-run matter in which the limitations of warships and fixed-wing aircraft in anti-submarine warfare (ASW) had been cruelly exposed. Submarine technology was little different from that which had prevailed in World War I, and the Allies had finally won as a result of superior tactical thinking, larger numbers and the eventual deployment of advanced technological developments such as radar, sonar and magnetic anomaly detection (MAD) gear. (The last was still rudimentary at the end of World War II, but proved valuable in the detection of submerged submarines by its ability to detect the distortions of the Earth's magnetic field caused by the localised mass of a submarine.) But the likelihood of nuclear-powered submarines from the 1950s onward elevated the threat to another dimension: such boats would be quieter and faster, posing distinct problems in themselves, and would have virtually unlimited underwater endurance, their nuclear reactors doing away with the need for rechargeable-battery technology and providing the power to recycle air or extract it from the water outside the boat. To this extent, therefore, the ASW problem of the 1950s was that of countering true submarines rather than submersibles, which is what World War II boats really were.

Warships and fixed-wing aircraft were still powerful weapons in ASW, but the warships were expensive and vulnerable, and the fixed-wing aircraft (both land- and carrier-based) were too fast. An ideal contribution could be made by the helicopter, however. This rotary-wing aircraft could operate from all types of vessel to which the submarine posed a threat, possessed sufficient speed to reach an operational area quickly, could hover while the sonar equipment was deployed, and the submarine detected and plotted. The helicopter could then attack the submarine with torpedoes or depth charges, or call in an accompanying warship for heavier support. Or so the theory went; although it was a fine theory, in practice it was hindered by the very limited payload of the helicopter, which precluded the carriage of sensors and weapons. The problem was exacerbated by the non-availability of dunking sonar, in which a powerful sonar transducer could be lowered from the belly of the helicopter to listen underwater for a submarine, then reeled up as the helicopter moved to a new location to dunk again and so secure the minimum of two bearings for a precise fix of the submarine's location. Early dunking sonars were limited in capability, largely as a result of their slow scanning in azimuth, so the tendency was to use sonobuoys which were light and could be scattered widely to secure a swift plot. But such expendable sonobuoys were expensive and not as powerful as dunking sonar, and the need to carry a relatively large number for any mission made the carriage of weapons impossible with the technology of the 1940s and 1950s.

No-one could deny the usefulness of helicopters for ASW or that ASW technology would take considerable strides forward during the 1950s. The first service to accept this fact was the US Navy, and in July 1950 Bell won the world's first design competition for an ASW helicopter. The resultant Bell HSL-1 was an oddity among Bell helicopters, being a tandem-rotor machine powered by a single 2,400-hp (1,790-kW) Pratt & Whitney R-2800 radial piston engine in the rear fuselage. Payload was an impressive 4,000 lb (1,814 kg) including electronic tracking gear, dunking sonar and weapons that included bombs, depth charges or the cumbersome Fairchild AUM-N-2 Petrel air-to-underwater missile. This last was essentially a turbojet-powered winged platform able to deliver a homing torpedo into the water some distance from the launch craft, but could only be carried by the HSL-1 when other equipment was omitted, its launch weight being 3,800 lb (1,724 kg) including the 2,000-lb (907-kg) torpedo. In historical terms, however, the HSL-1 worked as an effective ASW platform, validating the concept if nothing else. The trouble with the type, of which a mere 50 were built, was that it was too large and cumbersome for effective shipboard deployment, and had only a single engine with all its problems of reliability. Deliveries of the HSL-1 began in January 1957, and the type was used mainly for training.

Smaller helicopters could of course be produced, and in April 1950 the US Navy had contracted for examples of the Sikorsky S-55 single-rotor design under the designation HO4S-1. Here the limitations of the smaller helicopter become apparent, for the 'O' in the designation indicates that this S-55 derivative was an observation type: it was too small and limited in payload to carry sensors or weapons. Valuable operating experience was gained, however, and in June 1952 the US Navy placed an order for an ASW derivative of the Sikorsky S-58 under the service designation HSS-1, the middle 'S' of the designation indicating the type's role as being ASW. Though rotor diameter was increased by only 3 ft (0.91 m) in comparison with the 53 ft (16.15 m) of the HO4S series, power was more than doubled to 1,525 hp (1,138 kW), in this instance provided by a Wright R-1820 Cyclone radial. This large piston engine was located in the extreme nose at an oblique angle, and its long transmission shaft ran straight up between the pilots to the gearbox under the rotor hub. Payload was much increased compared with that of the HO4S, but the HSS-1 (invariably called Hiss-1 in service) still could not carry a useful combination of sensors and weapons. It therefore became standard practice to fit out the helicopters to operate as pairs, one carrying the dunking sonar and the other weapons. But the US Navy did not really favour such 'hunter/killer' helicopter teams, and the general rule was for the search helicopter to call in a destroyer for the submarine

An SH-3A Sea King of US Navy Squadron HS-11 hovers in the dunking mode.

'kill'. The type was a move in the right direction (particularly when the night-operation HSS-1N was introduced with automatic stabilisation equipment, Doppler navigation and an automatic hover coupler to keep the dunked sonar at the same depth and position regardless of the helicopter's movements): shipboard operation was facilitated by folding blades on the main rotor and a folding rear fuselage, but the limiting factor was still lack of payload and the relative lack of reliability from a single-engine powerplant.

The answer lay in turbine power, which offered fuel compatibility with the turbojets already almost universal in carrier-borne aircraft, plus increased mechanical simplicity (itself conducive to greater reliability), smoother running as a result of rotating rather than reciprocating movement of the most massive parts, and considerably reduced weight per unit of power delivered. General Electric had begun development of the T58 turboshaft in 1956 under naval contract, and on 30 January 1957 an HSS-1 modified to turboshaft power, under the designation HSS-1F, first flew with a pair of T58 turboshafts instead of the previous piston engine. The advantages were immediately obvious: each T58

was nearly as powerful as the R-1820 piston engine, yet weighed only 275 lb (125 kg) compared with the radial's 1,500 lb (680 kg). Thus a twin-turboshaft powerplant offered nearly twice the power of a single piston engine for slightly more than one-third the weight of the radial. Payload capability and reliability increased enormously, and the far smaller size of the turboshafts meant they could be removed from the nose of the airframe to a neat position with the gearbox under the rotor hub, leaving the main volume of the fuselage free for personnel and mission equipment. It was a quantum advance in helicopter development.

In 1957 the US Navy had outlined a requirement for a new ASW helicopter whose primary advance over older types was to be the combination into one airframe of the previously separate 'hunter' and 'killer' components of the complete ASW role. The US Navy therefore specified that dunking sonar was to be carried, that a weapon load of 840 lb (381 kg) was also to be accommodated, that fuel for a 4-hour mission was to be provided, and that equipment for night and all-weather operations was to be installed.

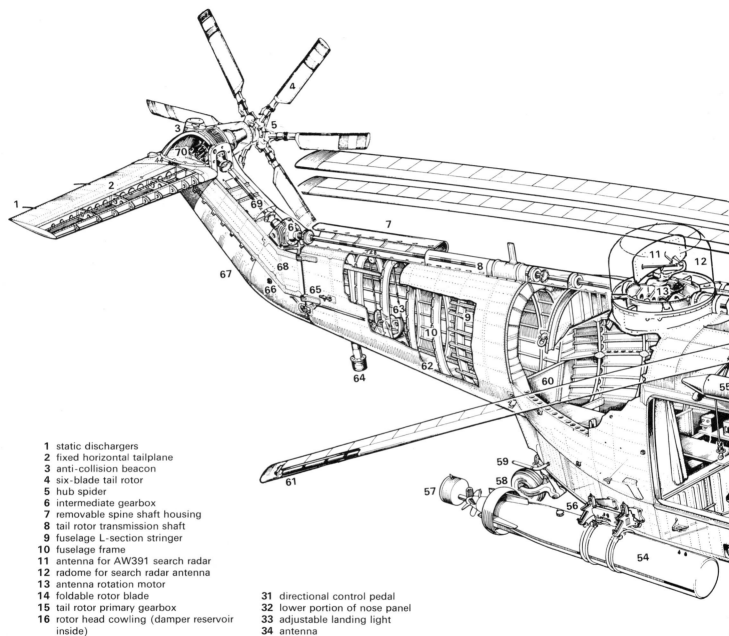

1 static dischargers
2 fixed horizontal tailplane
3 anti-collision beacon
4 six-blade tail rotor
5 hub spider
6 intermediate gearbox
7 removable spine shaft housing
8 tail rotor transmission shaft
9 fuselage L-section stringer
10 fuselage frame
11 antenna for AW391 search radar
12 radome for search radar antenna
13 antenna rotation motor
14 foldable rotor blade
15 tail rotor primary gearbox
16 rotor head cowling (damper reservoir
   inside)
17 blade root fitting
18 turbine exhaust
19 Rolls-Royce Gnome H.1400-1
   turboshaft
20 electric starter bullet as centrebody of
   turbine inlet
21 turbine cowling in open position
22 pitot head
23 upper portion of portside two-part
   airstair door
24 overhead console
25 windscreen washer/wiper unit
26 co-pilot's seat
27 nose hatch
28 hinged nose panel
29 fixed landing light
30 cyclic-pitch control column

31 directional control pedal
32 lower portion of nose panel
33 adjustable landing light
34 antenna
35 pilot's seat
36 cabin fresh-air inlet
37 handhold
38 sonar operator console
39 sponson strut
40 forward fuel cell
41 lash-down lug
42 sponson shock-absorber strut
43 access for sponson bilge pumps
44 navigation light
45 Plessey Type 195 dipping sonar
   transducer
46 twin-wheel main landing gear unit
47 leg of retractable main landing gear
   unit
48 transmission unit access panel in open
   position

49 air bottles
50 auxiliary flotation bag in stowed
   position
51 sliding freight door
52 aft fuel cell
53 pull-out emergency window
54 Mk 44 homing torpedo
55 Breeze BL 10300 variable-speed
   hydraulic rescue hoist
56 portside aft weapon launcher shackles
57 parachute unit
58 non-retractable tailwheel
59 fuel jettison pipe

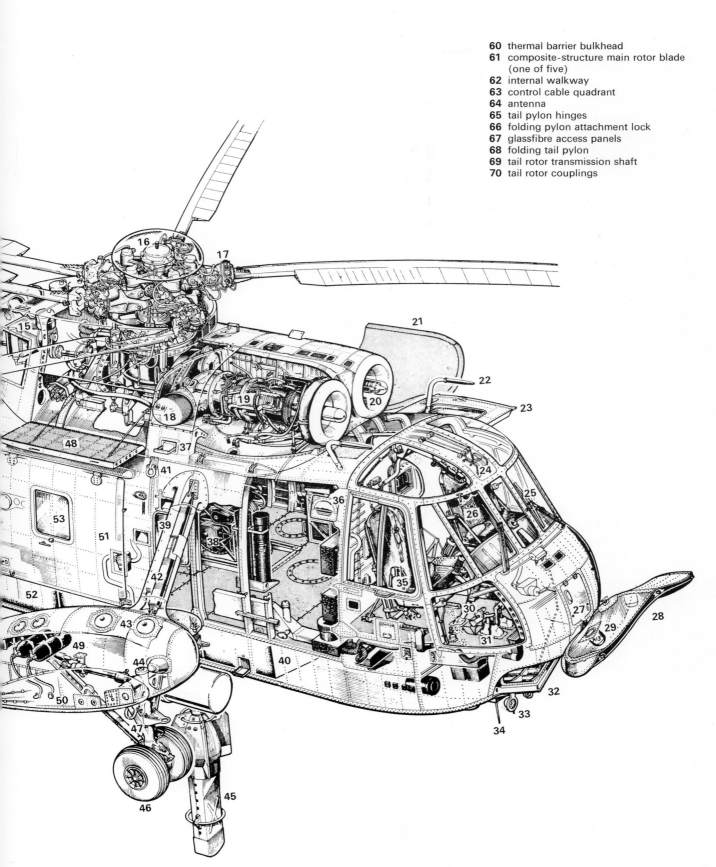

60 thermal barrier bulkhead
61 composite-structure main rotor blade
   (one of five)
62 internal walkway
63 control cable quadrant
64 antenna
65 tail pylon hinges
66 folding pylon attachment lock
67 glassfibre access panels
68 folding tail pylon
69 tail rotor transmission shaft
70 tail rotor couplings

Sikorsky S-61

Sikorsky's response was the **Sikorsky S-61**, which was recognisable as a development of the traditional Sikorsky design philosophy but considerably larger than the preceding HSS-1 and powered by a pair of T58 turboshafts. The development contract was signed on 24 December 1957, this calling for an initial service trials batch of seven **YHSS-2** helicopters. The S-61 had little in common with the S-58 (HSS-1) apart from conceptual similarities, but it suited the US Navy to imply (for financial and political reasons) that the new type was in fact a development, albeit radical, of a type already in widespread service. In September 1962 the US services adopted a unified designation system, and this resulted in the HSS-2's redesignation to **SH-3A**; so to avoid confusion, this later system is used hereafter as the primary designation.

As noted above, the S-61 was allied conceptually to the design practices that had prevailed for most previous Sikorsky helicopters: this dictated a pod-and-boom type of fuselage, single main rotor and single anti-torque tail rotor. However, the type's role as an overwater aircraft and the availability of the turboshaft powerplant meant that considerable detail refinement could be undertaken. Sikorsky's designers therefore evolved the S-61 as an amphibious design, the watertight boat hull permitting operations from the sea. The aircraft was balanced by lateral sponsons which also provided suitable housing for the main units of the retractable tailwheel type of wheel landing gear, permitting truly amphibious operation and thus enhanced operational flexibility. The lighter and more compact powerplant and its associated transmission system were located directly under the main rotor hub, so the designers could recast the main portion of the fuselage to optimise the crew's tactical capabilities. Thus the cockpit (laid out for two pilots) was moved down and forward into the extreme nose, forward of the main cabin which could accommodate the ASW tactical team and equipment or alternatively a load of freight or passengers. The fuselage itself followed standard Sikorsky practice, and was an all-metal flush-riveted structure of semi-monocoque design, with a fin-type swept rotor pylon having a fixed tailplane on the starboard side and the anti-torque rotor on the port side. The non-retractable tail wheel was attached to the extreme rear of the boat hull, just forward of the boom section. Much current work on helicopter improvements was concerned with rigid and other types of advanced rotor systems, but Sikorsky decided to stick with the conventional fully-articulated type using oil-lubricated bearings. The rotor hub was only slightly larger than that of the S-58 series, but held five rather than four blades. Each of these was of conventional structure, using an aluminium-alloy D-section leading-edge spar and honeycomb-filled trailing-edge pockets. To facilitate the type's deployment aboard ships, power-operated folding of the main rotor blades was provided, and the tail section (comprising the pylon, horizontal tailplane and five-blade anti-torque rotor) could be folded forwards by hand to reduce overall length.

Provision was made for armament to be carried under each of the two struts connecting the sponson stabilisers to the hull, up to a rated weight of 840 lb (381 kg). However, the normal load was a single 550-lb (249-kg) Mk 46 homing torpedo carried under the port strut, though nuclear depth charges could also be carried as an alternative. This provided the SH-3A with the required 'kill' capability but, as we have seen, US Navy operational doctrine has generally called for the helicopter to hunt the submarine and then call in a warship to effect the 'kill'. To this end the SH-3A has always been better provided with sensors than weapons. In the 1950s the US Navy's standard dunking sonar was the AN/AQS-4, a heavy and limited equipment whose search beam was only 10° wide. This meant that a full 360° search could take up to 5 minutes, with the possibility of a nuclear-powered submarine's astute commander evading the search or moving out of range before detection. The SH-3A was provided with a specially developed and considerably more advanced piece of equipment, the Bendix AN/AQS-10 with a 180° beam width and hence the ability to scan through 360° in only 1 minute. This equipment was in itself a great advantage, but its capabilities could only be used if the submerged transducer were held perfectly steady in the water regardless of the evolutions of the helicopter, which had ideally to hover away from the dunked transducer to avoid the extraneous noises resulting from rotor downwash. The SH-3A was thus provided with an automatic transition system able to bring the helicopter down from an altitude of several hundreds of feet to a stationary hover at an altitude of 50 ft (15 m), and with the aid of a sonar coupler (operating in conjunction with a radar altimeter and Ryan AN/APN-130 Doppler navigation radar) to hold the helicopter motionless relative to the dunked transducer. All-weather and night capability was ensured by the installation of a Hamilton Standard automatic stabilisation system.

Above: The gearbox and rotor head assembly of the Sea King series is notable for its sturdiness and small size.
Right: Essential maintenance is carried out on a six-barrel 7.62-mm (0.3-in) Minigun, one of the optional armament fits for several Sea King variants.
Below: Trial installation of a Minigun on the company-owned S-61 Interim AAFSS.
Below right: Main landing gear leg of an S-61 series amphibious helicopter.

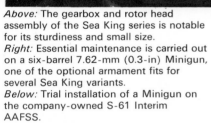

N318Y was the company-owned S-61A used for large numbers of trial installations. Notable here are the rescue winch on the starboard side of the fuselage, and also the stores attachment point under the strut supporting the starboard stabiliser sponson away from the fuselage. Extensive glazing gives the pilots exceptional fields of vision.

Into Service

The first YSH-3A (YHSS-2) flew on 11 March 1959, and there followed a two-year period in which the type was thoroughly evaluated by the manufacturer and the US Navy, a few problems being ironed out and the helicopter and its systems being validated as an effective ASW platform. This initial service version was designated **S-61B** by the company, the S-61A designation having been reserved for the proposed transport variant. (It is worth noting that the earlier S-58 series, which received the H-34 basic designation in October 1954, was also produced in a transport version as the widely used HUS-1, later UH-34D.) In February 1961 the programme was sufficiently advanced for the US Navy Board of Inspection and Survey to begin its service acceptance trials, and from May of the same year early production SH-3A Sea King helicopters were deployed to the Fleet Introduction Program, largely operated by US Navy Squadron VHS-1 based at the Naval Air Station (NAS) Key West in Florida. Conversion to the new type moved ahead smoothly, and the first two squadrons to become operational with the Sea King were VHS-3 at NAS Norfolk in Virginia, and VHS-10 at NAS Ream Field at San Diego, California. VHS-1 received its first SH-3As in May 1961, and the other two units received theirs in September of the same year. The advanced nature of the US Navy's new helicopter was demonstrated shortly after VHS-3 and VHS-10 became operational, when a Sikorsky-operated Sea King took the world helicopter speed record past the 200-mph (322-km/h) mark for the first time with a speed of 210.6 mph (339 km/h).

SH-3A Variants

The early **SH-3A Sea King** helicopters were powered by two 1,050-hp (783-kW) T58-GE-6 turboshafts, and the maximum take-off weight was limited to 18,000 lb (8,165 kg) with this powerplant. But General Electric was quick to improve upon the power output of this early turboshaft, at the same time improving the engine's reliability and responsiveness. Thus from the twentieth aircraft onwards these early Sea Kings were fitted with two 1,250-hp (933-kW) T58-GE-8 turboshafts, the availability of an additional 400 hp (298 kW) permitting an increase in maximum take-off weight to 19,100 lb (8,664 kg). Further improvement to the basic engine resulted in the designations T58-GE-8B and T58-GE-8F, but these improved the type's operational capability without adding to power output. The fairly rapid conversion of the US Navy's ASW helicopter squadrons to the Sea King

resulted in a steady stream of orders for the SH-3A, of which 245 were eventually built in batches dictated by the US fiscal year budget appropriations for US Navy aircraft.

The designation **CH-3B** came about in October 1962 for six examples of the Sea King operated by the US Air Force. Lacking any suitable helicopter of its own, and impressed with the payload/range performance of the US Navy HSS-2, the US Air Force in April 1962 borrowed three HSS-2s to supply 'Texas Tower' radar platforms in the Atlantic ocean. Required to operate from Otis AFB to the towers, the HSS-2s were stripped of ASW equipment to expose a cabin able to accommodate 27 people or up to 5,000 lb (2,268 kg) of freight. So successful was this conversion, which Sikorsky designated S-61A as the precursor of its transport version, that the USAF later in 1962 acquired three more examples of what was by then designated CH-3B under the tri-service designation system. The helicopters were later used

for missile-silo support duties and for the recovery of remotely-piloted vehicles (RPVs or drones).

The success of the S-61A in this USAF deployment confirmed Sikorsky's hopes for export orders, which soon began to come in to the company's New England factory in useful numbers. The S-61A series entered production to meet orders from the Royal Danish and Malaysian air forces. The former took nine S-61As optimised for the long-range air-sea rescue role but also able to operate with a load of 26 troops, or 15 litters in the medevac role, or 12 VIP passengers. Malaysia received 38 examples of the **S-61A-4 Nuri** up to 1978: these are multi-capable aircraft with auxiliary fuel tanks, a rescue hoist and seating for up to 31 passengers. A tenth S-61A went to a civil operator, Construction Helicopters.

Canada was also an early customer for the SH-3A variant, as attested by the original designation **CHSS-2** allocated to the helicopters ordered for delivery from May 1963. Intended for operations from Canadian warships, the CHSS-2 was later redesignated **CH-124**. While the first four were delivered from Sikorsky, the remaining 37 were built by United Aircraft of Canada, the local subsidiary of the United Technologies Corporation which also owns Sikorsky and Hamilton Standard.

Production of the S-61B for the US Navy totalled 265: 255 of these were of the main SH-3A variant, and the other 10 were ordered in 1961 under the designation **HSS-2Z** (altered to **VH-3A** in 1962). These 10 aircraft were allocated to the Executive Flight Detachment in Washington DC as Presidential and VIP transports, five aircraft

An SH-3A of US Navy Squadron HS-6 lifts off from the assault ship USS *New Orleans* during an Apollo space capsule recovery mission in July 1975.

An SH-3A (an aircraft built as part of the final production batch) on the strength of US Navy Squadron HS-6 flies a patrol with its MAD equipment deployed. The location is off San Clemente Island and the date 21 January 1976.

each being operated by the US Marine Corps and US Army as part of the tri-service EFD. The VH-3As are powered by T58-GE-8E engines, and are fitted with a high level of sound-proofing and extremely comfort-able interiors. One aircraft similar to the VH-3A was also produced for Indonesia, where it is operated by the department responsible for weapons main-tenance.

ASW continued to be the primary operational function of the Sea King in its SH-3A form, but the versatility of the basic aircraft recommended its use for a number of trials purposes. The first of these came in 1964, when nine SH-3As were modified into **RH-3A** mine-countermeasures helicopters, with special equip-ment for the sweeping of moored mines. Trials with such a method of sweeping (which kept the manned vehicle well away from the site of any possible ex-plosion) had been conducted with earlier types during the 1950s; but these had proved the

impracticality of piston-engined helicopters, which lacked the power to stream, tow and then recover the sweeping gear, which had first to be streamed by a surface vessel and then picked up for the tow by the helicopter. However, the RH-3A was capable of performing all three parts of the mission, albeit with so little a margin for error or malfunction that it was decided that any operational mine-countermeasures would have to have yet more power. The pro-gramme eventually provided the US Navy with the great Sikorsky RH-53D Sea Stallion, based on the S-65 with a total of 8,760 hp (6,535 kW) from two T64-GE-415 turboshafts. The squadron formed to conduct the RH-3A trials was VHM-12, and it is this unit which currently operates the RH-53D.

In 1970, some six years after the beginning of the RH-3A programme, the US Navy saw the need for an armed rescue helicopter able to penetrate combat zones of Vietnam to retrieve US Navy and US Marine Corps flight personnel from shot-down aircraft. The well-proved SH-3A was available for con-version, and Sikorsky produced kits for US Navy personnel at

Quonset Point to produce 12 **HH-3A** tactical search and rescue helicopters. The HH-3As were each powered by T58-GE-8F engines and provided with extra fuel capacity (fed by means of the 'Highdrink' refuelling system) and a high-speed rescue hoist. Just as significantly, however, each was provided with armour protection for crew and vital systems areas, and armament in the form of two 7.62-mm (0.3-in) Emerson Electric TAT-102 Mini-gun turrets. One was located in the rear of each sponson, where it was remotely controlled by a member of the crew in the cabin, or from the cockpit. The last aspects of the conversion were the plating-over of the vulnerable sonar well, and the reinforcement of the cabin floor. The first HH-3As were shipped during 1970 to US Navy Squadron VHC-7, based at Cubi Point in the Philippines, whence it moved up to Vietnam during 1971. Operating from a variety of bases and ships, the HH-3As proved an invaluable asset to the American naval forces in that dismal war, rescuing large numbers of downed personnel, and also boosting the morale of those who had to fly combat missions into inhospitable areas.

SH-3D Variants

By 1966 Sikorsky was able to offer a significantly improved version of the Sea King to the US Navy. The type had already been in service for nearly five years, and while a number of detail improvements and superior powerplants had been introduced on special variants, no major updating of the baseline SH-3A had been undertaken. Thus there appeared the **SH-3D**, of which Sikorsky built 73 for the US Navy, as the final new-build ASW variant for that service. The major change in the SH-3D was the introduction of a much improved powerplant comprising a pair of 1,400-hp (1,044-kW) General Electric T58-GE-10 turboshafts and a gearbox able to handle up to 2,500 hp (1,865 kW). The provision of such a gearbox was an added safety feature, allowing the remaining engine to develop its full military rating in the event of the other engine failing. The other features of the SH-3D were largely internal, and concerned the helicopter's tactical and avionics systems. An extra fuel tank and structural strengthening to increase maximum take-off weight to 20,500 lb (9,299 kg) were provided, and the instrument layout in the cockpit was improved. Most important, however, were the alterations to the electronics suite, where the improved AN/AQS-13A sonar replaced the original AN/AQS-13; AN/APN-182 Doppler radar supplanted the AN/APN-180 type; and an AN/ASN-50 heading reference system was added. Though small enough in themselves, these modifications as a package increased the combat effectiveness of the SH-3D quite markedly in comparison to that of the SH-3A. Six SH-3Ds were later modified to a standard similar to that of the earlier HH-3A.

Eleven examples of the **VH-3D** VIP/Presidential transport were also built as replacements for the VH-3As operated by the Executive Flight Detachment. High standards of soundproofing and comfort are again well in evidence.

In 1970 there appeared another derivative, the **SH-3G** based on the SH-3A. Sikorsky produced 105 SH-3G conversions by May 1972, the object being to provide the US Navy with a more versatile utility transport, but without losing the ASW capability of the SH-3A. Thus the main difference between the SH-3A and the SH-3G was the removability of the AN/AQS-13 sonar equipment. With this taken out of the cabin, it was possible to provide seated accommodation for 15; but the equipment was stowed on the helicopter, and it was a relatively simple task to restore it to the normal position and so return the SH-3G to ASW configuration. Most of the SH-3G conversions were on the strength of US Navy Squadron VHS-1, operating as a large number of small detachments on board US warships and at US bases all over the world.

During 1970 the latest ASW version of the Sea King was also under design in the form of the **SH-3H** variant. This resulted from a high-level requirement for a shipboard helicopter better able to tackle not only the new generation of faster, quieter and deeper-running Soviet nuclear-powered submarines, but also the sea-skimming anti-ship missile coming into prominence. Work on the SH-3H was thus centred in four main areas: retention and improvement of the convertibility features of the SH-3G, 12 improvements to the basic helicopter, six extra or improved

Opposite: With its automatic hovering coupler in operation, a Sea King of the US Navy dunks its sonar for the detection and bearing of any possible submarine in the search area.

items to the ASW suite, and new equipment to suit the SH-3H to the anti-shipping missile detection role.

All the convertibility features of the SH-3G were built into the SH-3H (a standard to which most surviving SH-3A, SH-3D and SH-3G Sea Kings were brought in a large-scale Sikorsky conversion programme that started in mid-1971 and produced its first fruit in spring 1972), and the SH-3H also has the ability to carry a pair of 110-US gal (418-litre) external fuel tanks in place of Mk 46 torpedoes. Other aids to survival over a modern naval battle arena are also provided for the SH-3H, including a Model H-240 chaff dispenser in a box on the port side of the fuselage between the rear of the sponson and the tail-wheel, and IDF-7.5 electronic surveillance measurement gear on the underside of the boom just aft of the tailwheel.

Aircraft improvements include standardisation on the T58-GE-10 turboshafts of the SH-3D, together with the uprated gearbox; structural modifications, largely to the sponsons and their struts, to permit an increase in maximum take-off weight to 21,000 lb (9,527 kg); the improved Doppler radar, the

heading reference system, the better IFF equipment and the secure voice transmission (by means of KY28 Juliet UHF radio) of the SH-3D; the increased fuel dump rate of the HH-3A; the improved automatic flight control system of the VH-3A; the cruise guide system (based on blade-stall indication) of the CH-53D and SH-3D; and an extra cabin window on the forward port side of the fuselage, a feature found in some SH-3As.

ASW improvements built into the converted SH-3Hs included the use of AN/AQS-13B Miniscan dunking sonar; the installation of AN/ASQ-81 MAD gear in the starboard sponson, with its towed 'bird' under the sponson in the stowed position; the ability to pre-set torpedoes in the air; the capability of launching torpedoes with the helicopter at the hover, thanks to the use of the Mk 31 Model O parachute-stabilised lanyard torpedo-delivery system; the fitting of 12 Type A sonobuoy launch tubes in the rear of the cabin, as already pioneered in some SH-3As; and the fitting of 24 Mk 25 marine-

Below: The quality of the paintwork gives a fair indication of the high standards required for the VH-3D Presidential transport, of which an example is seen under construction.

marker launch tubes in the rear of the port-side sponson, also as already pioneered in some SH-3As.

Missile detection was the task for which LN-66HP surveillance radar was provided, with its antenna in a 5-ft (1.52-m) radome under the boat hull section of the fuselage, providing the US Navy with a very useful warning system for the possible use of sea-skimming missiles against its powerful but also vulnerable major surface com-

batants. The SH-3H programme lasted up to 1980, and when it ended some 163 earlier Sea Kings had been brought up to this impressive multi-role standard. The only later US Navy designation is **YSH-3J**, which was allocated to two Sea Kings used by the Naval Weapons Test Center for the evaluation of weapons systems being con- sidered for the LAMPS (Light Airborne Multi-Purpose System) helicopter programme. Finally, it is worth noting that all Sea Kings

for the US Navy were built under the Sikorsky overall designation S-61B.

The story of the Sikorsky-built Sea Kings is completed by its few export versions, namely four SH-3Ds delivered to the Brazilian navy, four other helicopters to SH-3D standard delivered to the Argentine navy as **S-61D-4** air- craft, and 22 SH-3Ds for the Spanish navy.

Above: The third production SH-3A (then designated HSS-2) undergoes cold- weather tests. One of the few problems with the type has been ice ingestion, a problem cured by simple deflector shields in front of the inlets.
Left: Flightdeck crew complete the final preparations before a sortie from a carrier by this SH-3A of US Navy Squadron HS-11.
Below: In pristine condition, the first YSH-3A prototype (then designated YHSS-2) shows off its lines during the early flight test programme. Service demands for ever increasing amounts of extra equipment then began to mar the relatively clean lines of the basic aircraft, even the stabiliser floats being pressed into service as containers for essential gear.

The Westland Sea King HAS. Mk 1 was the Royal Navy's equivalent of the US Navy's SH-3D, but featured a considerably more comprehensive avionics fit, suiting the type to independent hunter/killer operations with a crew unaltered in number at four.
Inset: A Westland Sea King HAS. Mk 1 of No. 706 Squadron (based at RNAS Culdrose) dunks its Plessey Type 195 sonar transducer with the helicopter under the control of its Newmark Mk 31 automatic flight control system.

Right: A Royal Navy Sea King is helped down onto the deck of a British carrier.
Left: One of the Sikorsky S-61N's main roles has been in support of the expanding resources-exploitation industry, especially in the offshore oil and gas fields. Here an S-61N brings in a fresh crew for an offshore oil rig.
Below: Troops disembark from a Sea King HAS. Mk 5 helicopter onto the deck of a British aircraft-carrier. Though intended as an ASW platform, the Sea King HAS. Mk 5 has its general utility improved by its ability to carry passengers in emergencies.

Right: The Sea King is a fairly large heli-copter for shipboard deployment, but it can nevertheless use fairly small platforms, as indicated by this illustration of a Sea King HAS. Mk 1 of No. 826 Squadron, normally embarked on HMS *Bulwark* but here seen aboard a much smaller vessel. Note the dorsal radome for the Ekco AW391 surveillance radar.

Below: The best air-sea rescue helicopter available in the UK is the Sea King HAR. Mk 3, of which 16 are in service with the RAF's No. 202 Squadron, whose four-helicopter flights are deployed round the coast of the UK. Note the bulged rear cabin windows. Accommodation is provided in the cabin for 19 seated survivors, or 11 seated survivors and two litters, or six litters.

Left: A Sea King HC. Mk 4 shows off its ability to carry a substantial slung load, one of the factors that made the type so useful in the Falklands campaign, in which 105-mm howitzers and their ammunition were carried.

Below: Carrying temporary British markings, one of the six Sea King Mk 47 ASW helicopters for Egypt shows off its paces before delivery in 1976. Payment for the helicopters was made by Saudi Arabia.

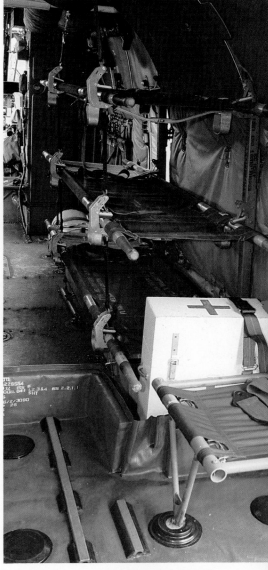

Above: Again sporting a temporary British civil registration, this is a Sea King Mk 45 for the Pakistani navy, one of six whose delivery was completed in 1975.

Right: The roomy cabin of the Sea King series makes it relatively simple to instal six litters, with ample room left for the attendants and cabin crew.

Below: The flightdeck of the Sea King is well designed along ergonomic lines for the pilot and co-pilot in right and left seats respectively.

The Sea King Mk 48 is the SAR version for the Belgian air force, whose five examples (including one with VIP interior capability) are stationed at Coxyde with No. 40 Squadron.

Inset: Westland's first export customer for the Sea King was West Germany, which ordered 22 Sea King Mk 41 SAR helicopters. From 1984 the 20 survivors will be upgraded with a measure of combat capability.

Above: The Westland Commando is structurally very close to the Sea King apart from its revised landing gear, but the type carried an extremely wide variety of armament options. Seen here are Commando Mk 2s of the Egyptian air force before delivery.

The Commando Mk 2E is a specially adapted version of the basic helicopter for electronic warfare duties. The type was developed to meet the particular requirements of Egypt.

Above: Sea Kings ferry supplies ashore during the Falklands campaign, during which the type displayed excellent reliability and great versatility under the most arduous of conditions.

Below: One of the most important lessons of the Falklands campaign was the need for an airborne early warning aircraft, and the hastily improvised but wholly successful Sea King AEW resulted, with Thorn-EMI Searchwater radar in two converted Sea King HAS. Mk 2 airframes. It is likely that five Sea King HAR. Mk 3 helicopters will receive similar conversions.

HH-3E Jolly Green Giant

During the early 1960s the US Air Force began to formulate a requirement for a long-range medium-lift transport helicopter, and its experience with the HSS-2s operated as CH-3B interim transport helicopters persuaded the USAF that the ideal type could be provided by Sikorsky. As noted above, the HSS-2s for the USAF had been modified by Sikorsky into a configuration designated S-61A by the company, and other helicopters to similar standards had been exported to Denmark and Malaysia. But little more was achieved with this minimum-change transport version of the S-61B, and the USAF requirement called for an aircraft with a number of major modifications, which Sikorsky designed under the designation **S-61R**. The dynamic system and forward fuselage remained unaltered in comparison with those of the S-61B, but to facilitate USAF operations all features concerned with amphibious and ASW capability were removed, and the rest of the airframe was modified to suit land-based tactical transport use. Here the company was greatly aided by design work undertaken during 1959 in an effort to meet a US Marine Corps requirement for a medium transport helicopter. Sikorsky had proposed a variant of the S-61 under the official designation **HR3S**, but this failed to secure backing. However, much of the effort bore fruit in the rapid evolution of the S-61R, whose main features were test flown in a company-owned aircraft (registered N664Y) that first took to the air on 17 June 1963. By this time the USAF had contracted for 22 production examples, to be designated **CH-3C**, and this letter contract of 8 February 1963 was later

supplemented by others raising total procurement of the CH-3C to 75.

The fuselage of the CH-3C was considerably revised aft of the cockpit in comparison with that of the Sea King series. The main cabin was extended aft to provide an internal length of 25 ft 10½ in (7.89 m), and though cabin width remained unaltered at 6 ft 6 in (1.98 m) and height only very slightly reduced from 6 ft 3½ in (1.92 m) to 6 ft 3 in (1.91 m), the usable floor area was increased to 168 sq ft (15.61 m²) and volume

Above: Development of the S-61R series for the US Air Force was undertaken with the Sikorsky-owned N664Y. Differences from the original Sea Kings are immediately apparent: substitution of rear stubs for the original stabilising sponsons, use of tricycle landing gear, and the alteration of the rear fuselage to permit the installation of a ramp door. Also noteworthy is the larger, braced horizontal tailplane.
Below: N664Y is pictured during evaluation by the USAF, which procured the type under the designation CH-3C, the first of a relatively small but very significant series including the classic HH-3E Jolly Green Giant rescue helicopter.

to 1,050 cu ft (29.73 m³). Just as importantly, the sharp edged rear to the boat hull of the Sea King was removed to allow the provision of a full-width rear ramp/door, hydraulically operated to permit the straight-in loading of freight and even light vehicles. An internal winch, rated at 2,000-lb (907-kg) capacity, was installed for the handling of freight within the cabin. Nominal payload within the cabin was 25 troops or 5,000 lb (2,268 kg) of freight.

To provide the height necessary for the ramp/door at the rear of the cabin, the boom supporting the tail unit and anti-torque rotor was reduced considerably in depth, and another useful visual identification feature of the CH-3C series was the revised horizontal tail, which was a strut-braced unit spanning 9 ft 5 in (2.87 m). The tailplane of the Sea King series was 6 ft (1.83 m) in span and unbraced.

Another notable alteration in the CH-3C was the landing gear. The strutted sponsons of the Sea King were replaced by sponson stub wings in a position farther aft, and the conversion from the semi-retractable tailwheel landing gear of the Sea King to the fully retractable tricycle landing gear of the CH-3C was completed by the addition of a twin-wheel nose leg under the forward portion of the cabin. As in the Sea King, actuation of the landing gear was hydraulic, and the retention of

the sponsons (albeit in much-altered form) gave the CH-3C a great deal of buoyancy (9,594 lb/4,352 kg) as a safety factor in the event of a ditching. Detail alterations, hardly visible to exterior detection, also added significantly to the operational capability of the CH-3C. These alterations were designed to permit the deployment of the type away from base servicing facilities (a factor unimportant with the Sea King, which was intended for operations from warships and bases with good technical resources), and included pressurised rotor blades for quick and easy inspection for cracking, self-lubricating main and anti-torque rotors, inbuilt equipment for the maintenance, removal and replacement of major components in the field, and a gas turbine auxiliary power unit for self-start capability in independent operations.

The first CH-3C was delivered to the USAF on 30 December 1963, and the type was soon operational at Tyndall AFB in Florida for drone recovery duties. The 75 production examples of the CH-3C, all powered by two 1,300-hp (970-kW) General Electric T56-GE-1 turboshafts with a total fuel capacity of 642 US gal (2,430 litres) in a pair of bladder tanks under the cabin floor, were soon in widespread service with the Strategic Air Command, Tactical Air Command, Aerospace Defense Command,

Air Training Command and Aerospace Rescue and Recovery Service. There could be no better proof of the CH-3C's reliability and versatility than the fact that so few of the type performed so creditably and constantly with five major elements of the USAF.

In February 1966 there appeared the **CH-3E**, an up-engined version of the CH-3C with 1,500-hp (1,119-kW) General Electric T58-GE-5 turboshafts (the USAF's equivalent of the US Navy's T58-GE-10) replacing the earlier T58-GE-1 engines (the USAF's equivalent of the US Navy's T58-GE-8). Some 42 new-build CH-3Es were produced, and all CH-3Cs were recycled through the Sikorsky plant for upgrading to CH-3E standard. Maximum cabin accommodation was increased to 30

By now US forces were heavily embroiled in the Vietnam War, and the USAF discovered a pressing need for a helicopter able to penetrate deep over hostile territory to recover the crews of downed aircraft. Evaluation of the concept of an armoured and armed S-61R variant was undertaken with a pair of CH-3Es, and in 1965 it was decided to modify some 50 CH-3Es to a much revised standard as **HH-3E** combat-area rescue helicopters. The revised type was quickly nicknamed **Jolly Green Giant**, and the success of the HH-3E was capped by official acceptance of the name. Modifications to the dynamic system and airframe of the CH-3E were minimal in the change to HH-3E standard. But the internal provisions of the aircraft and various systems were much altered: protective armour was provided for the crew and essential systems, self-sealing fuel tanks became standard, provision was made for jettisonable long-range tanks under the sponsons, a telescopic inflight-refuelling probe was added under the starboard side of the nose to permit the HH-3E to refuel from Lockheed HC-130P tankers, a high-speed rescue hoist was added, and allowance was made for defensive armament in the form of one 7.62-mm (0.3-in) Minigun or one 0.5-in (12.7-mm) Browning M2 heavy machine-gun firing through a cabin window on each side of the air-craft. Some internal modifications were made to the airframe, and this permitted an increase in maximum take-off weight to 22,050 lb (10,002 kg), to which CH-3Es were also cleared at a later date.

The HH-3E was clearly a very capable machine, and an indication of the type's potential was provided between 31 May and 1 June 1967, when two HH-3Es made the first nonstop helicopter crossing of the North Atlantic en route to the Paris Air Show. The helicopters refuelled in the air some nine times, and the flight of 4,270 miles (6,870 km) was achieved in 30 hours 46 minutes at an average speed of just under 139 mph (224 km/h). Despite the proven reliability of the airframe/twin-engine powerplant combination, it must have been of comfort to the crews involved that the HH-3Es, though without the boat hulls of the Sea King series, were nevertheless finished with water-tight hulls for just such a contingency. The HH-3E was allocated to squadrons of the USAF's Aerospace Rescue and Recovery Service, and these took the type to Vietnam during 1966. In that theatre the HH-3E soon built up an enviable reputation for the recovery of aircrew in distinctly hostile environments, including North Vietnam. The unit with the best success ratio was the 3rd Aerospace Rescue and Recovery Group, operating as part of the 7th Air Force from a main base at Tan Son Nhut Air Base outside Saigon, with detachments often dispersed to 17 Vietnamese and four Thai bases. The 3rd ARRG operated a mixed force of helicopters and Hercules tankers, and could call on tactical units for close support in hostile areas. The most frequently requested tactical air-craft was the Douglas A-1

An HH-3E comes in low, with a sting in its tail provided by a 7.62-mm (0.3-in) Minigun on the half-open ramp.

Skyraider, which had the fire-power and endurance to offer long support deep in hostile territory, but possessed a cruising speed similar to that of the HH-3E, so permitting a close escort to and from the pick-up area. The technique was for the downed aircrew to make for the nearest clear ground, preferably a hill top, and for the tactical air-craft to suppress the defences while the HH-3E swooped in to rescue the crew with either a landing or the 600-lb (272-kg) capacity winch, which had 240 ft (73 m) of line.

Even before the HH-3E had reached Vietnam, the US Coast Guard announced that it too was to buy a rescue helicopter based on the same basic concept. This was the **HH-3F Pelican**, of which 40 were ordered from August 1965, with deliveries beginning in 1968. The HH-3F is basically similar in the airframe and dynamic system to the HH-3E, but has no provision for protection or armament, and there is no inflight-refuelling probe. But from the Sea King type the HH-3F has acquired a boat hull to suit the helicopter for long-range search and rescue operations over the waters off the US east, west and south sea-boards. Intended for operations in all weathers, the HH-3F has advanced electronic systems and communications gear, including search radar with its antenna in a radome projecting from the nose under the co-pilot's windscreen. The HH-3F is configured for a crew of four (two pilots in the cockpit and the navigator and winch operator in the cabin) with provision for up to 15 stretcher patients.

The US Coast Guards' HH-3F Pelican is an unarmed derivative of the HH-3E with special rescue equipment and advanced avionics including radar.

Experimental Variants

Two of the major disadvantages of the pure helicopter in comparison with fixed-wing aircraft have always been the rotary-wing craft's lack of speed and high disc loading. The latter presents severe structural problems and inhibits manoeuvrability, both factors with possibly serious consequences for military operators. Much research work had been done into the problem during the 1950s and early 1960s, but the development of turbine-powered multi-role types such as the Sikorsky S-61 gave added impetus to the programme. These helicopters had ample margins of power for development purposes, were well proved in service, and had established a bank of operating data that made them excellent 'control' aircraft against which the performance of the experimental models could be compared. For Sikorsky this research programme began to look hopeful in July 1964, when the company received a joint US Army/US Navy contract for the modification of an SH-3A into the **NH-3A** (company designation **S-61F**) high-speed research helicopter. The modifications needed for the dynamic system were minimal, but the airframe was more extensively altered, though mostly in cosmetic aspects to improve streamlining: the boat hull was replaced by a rounded bottom that continued into the under surface of a deepened boom; the stabilising sponsons were removed, together with their struts; the tail unit was enlarged, a symmetrical tailplane was fitted and a six-blade anti-torque rotor was provided; and the landing gear was revised to feature main units retracting fully into the lower fuselage and a fixed tailwheel under the rotor pylon. So far the NH-3A was a revision of the standard SH-3A, but a completely different look was given to the machine by the extra powerplant, comprising a pair of 3,000-lb (1,361-kg) thrust Pratt & Whitney J60 turbojets, mounted in pods located at the tip of the stub wings projecting from the sides of the lower fuselage. In this form the NH-3A first flew on 21 May 1965, and flight trials revealed a maximum speed of 242 mph (390 km/h). After exhaustive trials with the NH-3A in this form, a helicopter with auxiliary power, the aircraft was revised to compound configuration with a pair of wings, spanning 32 ft (9.75 m), as a means of offloading the main rotor in forward flight.

A more radical redesign (or perhaps even half-sister) was the **S-67 Blackhawk**. This was a company-funded attempt to provide the US Army with an 'AH-3' gunship helicopter. The company had high hopes for the project, whose detailed design began in August 1969. The Sikorsky S-66 design had lost to the Lockheed CL-840 in the US Army's 1965 competition for an Advanced Aerial Fire-Support System, but the resultant AH-56A Cheyenne was by the later 1960s in serious flight and financial problems. Sikorsky saw the chance to step in with its interim design combining the dynamic system of the S-61 with features of the S-66 and the compound aspects of the S-61F. Construction of a prototype high-speed attack helicopter began in November 1969, and the first flight was made on 20 August 1970, completing its initial trials successfully only one month later. The dynamic system was modelled closely on that of the S-61R series, using the same T58-GE-5 turboshafts, five-blade main rotor and five-blade anti-torque tail rotor. However, the blades of the main rotor were swept back to delay blade-tip stall at high speed, to reduce vibration stresses and to improve the lift/drag ratio. The dynamic system was also made less 'draggy' by better fairing of the rotor head and more slippery air inlets for the side-by-side turboshafts. The fuselage was much reduced in width compared to that of the Sea King, to produce a flat-plane frontal area of only 17 sq ft (1.58 m^2) compared with 32 sq ft (2.97 m^2) for the SH-3A. Thus the two flightcrew were seated in tandem under a low canopy, the pilot in the front seat and the co-pilot/gunner behind and above him; the fuselage was nonetheless roomy, in the manner of the Russian Mil Mi-24 'Hind', and could provide accommodation for 15 troops on the upper 'deck' of the two-level rear compartment, whose lower level housed fuel and extensive quantities of ammunition (up to 1,500 rounds of 30-mm calibre). However, the most impressive feature of the S-67 was the armament. Only one system was inbuilt, this being a TATH 140 turret under the fuselage, with another turret optional in the nose position. The TATH 140 was a versatile mounting able to carry cannon of 20- or 30-mm calibre, or a 7.62-mm (0.3-in) Minigun, or a 40-mm grenade-launcher. Extra armament capability was provided by the optional wings, which spanned 27 ft 4 in (8.33 m) and were fitted outside the sponsons housing the main units of the retractable tailwheel landing gear. These wings provided six hardpoints for the carriage of 16 TOW anti-tank missiles or eight 19-tube rocket-launchers, to provide a maximum ordnance load of some 8,000 lb (3,629 kg) in a maximum take-off weight of 22,050 lb (10,002 kg). It was also proposed that the S-67 should have some defence capability against fighters by the provision of AIM-9 Sidewinder air-to-air missiles, one on each wingtip.

The performance of the S-67 was impressive: on 14 December 1970 Byron Graham flew the machine to a speed of 216.844 mph (348.971 km/h) over a 3-km (1.86-mile) straight course, and on 19 December 1970 Kurt Cannon flew the same aircraft over a 15/25-km (9.32/15.53-mile) circuit at 220.885 mph (355.485 km/h). Both speeds were ratified as world helicopter records. The helicopter was also very manoeuvrable, being capable of rolls and split-S turns without difficulty. In 1972 the S-67 was flown off against the Lockheed Cheyenne and the Bell Model 309 KingCobra. The Lockheed design was eliminated from the com-petition, but eventually the US Army decided that its current Advanced Attack Helicopter requirement could only be met by a new design. That finally selected for service in the early 1980s was the Hughes Model 77, in service designated AH-64A Apache and a very powerful, very complex and very expensive machine.

S-61 Airliners

From the beginning of the S-61 project Sikorsky had borne in mind the possibility of pure transport versions, and the first off the ground was the S-61A series of military transports. This was not a great success commercially, largely as a result of the fact that the only Western forces with a requirement for substantial VTOL airlift capability were the US Marine Corps and the US Army. The former had useful numbers of relatively new Sikorsky HUS-1 Seahorse helicopters in service and in the early 1960s had a requirement for a more advanced and versatile helicopter than could be met by the S-61 design. The latter was also the operator of an S-58 variant, the H-34 Choctaw, and the several hundreds of these provided the US Army with its required assault-helicopter force until the early 1960s, when a more versatile helicopter was again required. For military transport purposes, at least on a substantial production basis, the S-61 had arrived too soon.

Sikorsky's hopes were thus pinned on orders for airliner developments of the basic type. The company was all too aware of the limitations of the helicopter as a civil transport. As an airliner the helicopter was extremely noisy, slow and cost-ineffective in comparison with fixed-wing types: the S-61A military transport, for example, required 2,500 hp (1,865 kW) to move 26 passengers at about 150 mph (241 km/h); the contemporary Avro 748 fixed-wing transport

Opposite, above: Designed as a gunship based on the dynamic system of the S-61 series, the S-67 was aerodynamically clean and had a crew of two in tandem in the slim fuselage.
Below: Drop tanks do little to mar the lines of the S-67, but the undernose turret did reduce performance by a small margin.

could carry 48 passengers at some 280 mph (451 km/h) on a total of 3,480 hp (2,596 kW). The fixed-wing transport could, therefore, carry nearly twice as many passengers at nearly twice the speed on only 40 per cent more power. Moreover, the fixed-wing transport had great range advantage.

But Sikorsky pinned its hopes on the helicopter's unique capabilities of VTOL, which suited the type for passenger transport between city centres, airports and the like, and which also made the helicopter a valuable tool for specialist tasks such as logging, support of the offshore oil rigs that were becoming an important part of the world's energy-recovery programme, and certain construction tasks. Sikorsky accordingly began work on two basically civil derivatives of the S-61 family. The first to fly, on 6 December 1960, was the prototype **S-61L**, which was intended for operations over land. This was based on the SH-3A, using the same dynamic system and an airframe allied to the initial naval model. The airframe was considerably adapted to suit it to its new role, however. The boat hull was replaced by a more streamlined conventional fuselage, which was also 'stretched' by 4 ft 2 in (1.27 m) forward of the main rotor. Power was provided by a pair of 1,350-hp (1,007-kW) General Electric CT58-110 turboshafts, and the sponsoned landing gear of the naval version was replaced by a simple strut-mounted wheeled landing gear of tailwheel configuration. The hull was sealed as insurance against the possibility of an emergency landing on water, the cabin was furnished to airline standards for 28 passengers, and provision was made for a crew of three (two pilots and a cabin attendant). Facilities included a small galley and lavatory, and access to the cabin was provided on the starboard side by a pair of doors, of

which the rear was an airstair unit. An option provided for customers was the possibility of folding seats in the forward half of the cabin to permit the carriage of freight in this section. All S-61Ls were provided with full blind-flying instrumentation, with other avionics left to customer discretion.

The S-61L was certificated by the Federal Aviation Administration on 2 November 1961, and the type entered service with Los Angeles Airways in time for a first revenue flight on 1 March 1962. Los Angeles Airways was the world's first airline to operate scheduled passenger services with helicopters (in 1947), restricting itself to terminals within a 50-mile (80-km) radius of the post office in downtown Los Angeles, and so operating a unique network connecting the city, airports and areas of conurbation ideally suited to helicopter operations. However, the restricted suitability of the S-61L meant that total sales amounted to only 13 aircraft. This was in part attributable to the S-61L's suitability for a single role only, but also in part to the type's relatively small payload, as indicated by an empty weight of 11,704 lb (5,309 kg) and a maximum take-off weight of 19,000 lb (8,618 kg).

The second civil derivative of the S-61 family was the **S-61N**, an amphibious type resembling the Sea King more closely, thanks to the retention of the boat hull, stabilising sponsons and retractable landing gear. But the same type of fuselage 'stretch' was applied, with the result that both the S-61L and the S-61N have a cabin measuring 31 ft 11 in (9.73 m) in length, 6 ft 6 in (1.98 m) in width and 6 ft 3½ in (1.92 m) in height, with a floor area of 217 sq ft (20.16 m²) and a cabin volume of 1,305 cu ft (36.95 m³). The volume available for freight was 150 cu ft (4.25 m³), 125 cu ft (3.54 m³)

of this above the floor and the remainder below it. The first S-61N flew on 7 August 1962, and the type found a readier market for airline and utility operations. Like the S-61L, the early S-61N models were powered by CT58-110 turboshafts (the civil equivalent of the T58-GE-8), but from 1969 the S-61N became available in **S-61N Mk II** form with a number of improvements, including 1,500-hp (1,119-kW) CT58-140-2 turboshafts. The S-61L had been the first twin-turbine civil helicopter to receive

Opposite: This overhead view of the S-61L shows the simple arrangement of the powerplant and rotor system, and also the landplane arrangement of the main-wheels.

Below: The S-61N, whose first operator was Los Angeles Airways, combined features of the S-61L and the original Sea King, for example combining the Sea King's boat hull with the stretch introduced by the S-61L.

FAA certification and, on 6 October 1964, the S-61L and S-61N were each certificated for full Instrument Flight Rules (IFR) operations, again being the first civil transport helicopters to receive such certification. This capability was retained in the S-61N Mk II, which introduced weather radar as standard, had slightly enlarged sponsons, was fitted for the carriage of a few more passengers, and had improved anti-vibration measures to reduce the amount of vibration transmitted into the fuselage from the main rotor.

The third civil derivative of the S-61 series was the **Payloader**, based on the S-61N but fitted with the fixed landing gear of the S-61L and stripped of all non-essentials to increase payload to over 11,000 lb (4,990 kg). The airstair door was sealed, and the type was optimised for heavy-

lift work in such fields as construction, logging, pipeline laying and powerline installation.

Construction of the civil S-61 variants ended in June 1980 after the production of the 13 S-61Ls and 123 examples of the S-61N and Payloader. Though these are not very high figures for overall production, they nevertheless represented a considerable profit for Sikorsky, for most of the engineering and tooling had been produced under military contract, so these civil aircraft were produced at a higher profit margin. Throughout the 1970s the S-61N in particular was rightly regarded as the long-haul workhorse of the increasingly important offshore resources industry, the main operator of the type being British Airways, which at one time had no fewer than 23 S-61s, largely in support of the oilfields in the North Sea.

British Airways is currently the world's largest operator of the S-61N, this being one of 24 such helicopters in service for support mainly of the oil industry in the North Sea. Note the weather radar in the substantial thimble radome, without which North Sea operations would be out of the question.

Licence-Production in Italy and Japan

Not surprisingly, major nations with important maritime interests were eager to procure the Sea King for their navies, but only on terms advantageous to their indigenous aircraft industries. Smaller nations were happy to accept Sikorsky-built helicopters (as detailed above), but Italy, Japan and the UK saw licence production of the Sea King type in terms not only of providing the right kind of ASW capability, but also of acquiring advanced American helicopter technology while curtailing the expenditure of hard-earned US dollars.

Sikorsky concluded a licence-production agreement with Costruzioni Aeronautiche Giovanni Agusta SpA for the S-61, the areas in which Agusta could sell its Sea King models being fixed as Italy, North Africa and the Middle East. Production got under way in 1967 against an initial order for the Italian navy. Although the designation is not an official one, it may be convenient to categorise these Italian aircraft with the designatory prefix AS (Agusta-Sikorsky), analogous to the more official AB (Agusta-Bell), to differentiate the products of the Italian and American production lines. The Italian navy's order was for 24 **A-SH-3D** helicopters (**AS-61B**), and deliveries began in 1969. The type was essentially similar to the SH-3D apart from some local strengthening of the airframe to suit the AS-61B for a more multi-role capability, the installation of an improved tail-plane, the use of two 1,500-hp (1,119-kW) General Electric T58-GE-100 turboshafts, and the avionics and weapons fit. So far as the AS-61B's capabilities were concerned, the Italians wanted (and received) a lot for their money, the AS-61B being operable in several roles: ASW with full search, identification and attack capability thanks to the provision of dunking sonar, 360° search radar with its antenna in a ventral radome, a tactical compartment in the fuselage, armament of four homing torpedoes or depth charges and a host of other equipment; ASV (anti-surface vessel) warfare with search and command radar (AN/APN-195 in a nose mounting designed by Sikorsky for the HH-3F) operating through a special computer interface to feed data for the use of four AS.12 medium-range or two Sea Killer Mk 2/AM.39 Exocet/Harpoon long-range anti-ship missiles, even in an environment of heavy electronic countermeasures; SAR (search and rescue) with a variable-speed hydraulic winch (600-lb/272-kg capacity) above the sliding starboard-side door, and accommodation for 25 survivors and medical crew even with radar installed; medevac with accommodation for 15 litters and one medical attendant; and transport with internal accommodation for 31 troops carried over a range of 362 miles (582 km), or 6,000 lb (2,722 kg) of freight, or with an external load of 8,000 lb (3,629 kg). Eventual procurement of the SH-3D for the Italian navy was 27, and this production for the Italian services was supplemented by two **A-SH-3D/TS** (Trasporto Speciale) VIP transports for the Italian air force.

Also produced by Agusta was the **AS-61A-4** multi-role export version of the AS-61, suitable for trooping, freighting, SAR and medevac duties with a crew of three plus up to 31 troops, or 15 litters or 10 VIP passengers. In this last configuration, AS-61A-4s have been sold to Iran, Libya and Saudi Arabia. Both these AS-61 models can be brought up to SH-3H standards as required.

Also built by Agusta are the **AS-61R**, an Italian version of the HH-3F Pelican, of which 20 are

A powerful anti-submarine component of the Italian navy, the Agusta-Sikorsky SH-3D/H series in its SH-3H form as shown, adds effective anti-ship capability.

Above: The Agusta-Sikorsky S-61R is the Italian air force's equivalent of the US Coast Guard's HH-3F Pelican, a very versatile search-and-rescue helicopter.

Below: Combining retractable tricycle landing gear with a watertight boat hull, the Agusta-Sikorsky S-61R is fully amphibious in its SAR role.

An S-61N of Bristow Helicopters, a major
operator of the S-61N until the loss of
such a helicopter in a North Sea crash
following rotor failure.

operated for SAR duties by the Italian air force, and the **AS-61N-1 Silver** civil derivative, able to carry 23 passengers in airliner comfort over ranges of 645 miles (1,038 km).

Chronologically, Mitsubishi Jukogyo Kabushiki Kaisha (Mitsubishi Heavy Industries Ltd) of Japan was the first foreign licensee for the S-61 series to get into production, its agreement with Sikorsky dating from May 1962, when the Japanese Maritime Self-Defense Force ordered 11 examples of the HSS-2 (SH-3A). The first of these was delivered as a complete aircraft by Sikorsky, the second and third aircraft were delivered in partly-assembled form for completion by Mitsubishi, and the subsequent aircraft were built by Mitsubishi with a decreasing proportion of components supplied by Sikorsky. Mitsubishi holds licences for the S-61A and S-61B series, and the Japanese aircraft are identical with their American counterparts other than powerplant, which comprises a pair of Ishikawajima-Harima T58-IHI-10 turboshafts, the Japanese-built version of the T58-GE-10. Japanese production (here categorised by the prefix letters MS for convenience rather than strict accuracy) comprises 123 **MS-61B** and **MS-61B-1** ASW helicopters (generally regarded as uprated versions of the SH-3A) for the Japanese Maritime Self-Defense Force, plus 10 **MS-61A-1** SAR helicopters and a complement of one MS-61A and three MS-61A-1 helicopters for the Japanese Antarctic expedition.

Right: Cleverly designed on ergonomic principles, the instrument layout of the Sea King series is combined with good fields of vision for the flightcrew.
Below: The pilot's position in a Sea King.

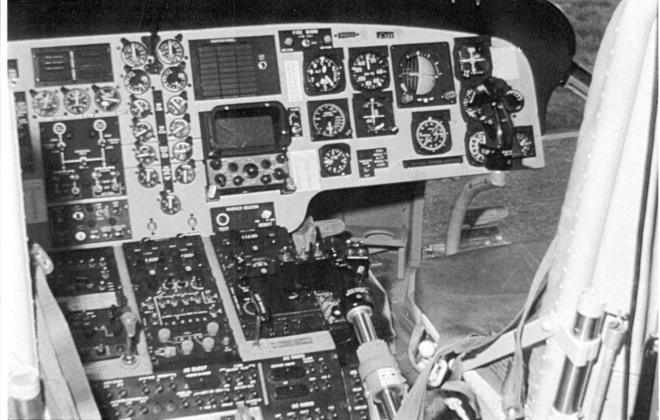

Westland Sea King

Sikorsky's longest-established licence agreement is with Westland Helicopters in the UK, the agreement dating back to the 1940s. But in the case of the Sea King, Westland signed a licence agreement with Sikorsky in 1959, though there was at the time little real thought of British production of the type, efforts by the Royal Navy to secure an advanced ASW helicopter being centred upon the Westland WG.1, a project derived from the Bristol Types 173 and 192 tandem-rotor helicopters. The all-too-familiar efforts of economy-minded government departments contrived to combine the Royal Navy's requirement for an ASW helicopter with the already entwined needs of the Royal Air Force and Army for a heavy-lift helicopter, the resultant delays ensuring that the possible outcome was too late for the Royal Navy's already pressing needs. At the urging of the Royal Navy, therefore, the government agreed to a purchase of Sea Kings. In its effort to sell the S-61 in the UK, Sikorsky tendered for US-built aircraft, with the proviso that a specific licence-production agreement would be signed if Westland could underbid the parent company. This latter occurred, and on 27 June 1966 Westland was awarded a contract for 60 Sea King helicopters.

Unlike the Italian and Japanese versions of the Sea King, these British machines were considerably altered compared with the original American helicopter. Earlier production by Westland of Sikorsky designs (notably the S-55 and S-58) had been of a fully 'Anglicized' product, but in the case of the S-61 Westland worked to Sikorsky blueprints with freedom to make changes wherever necessary. However, so urgent was the Royal Navy's

requirement that many items of equipment standard to the SH-3D model selected as the British basic type were retained to speed the development process.

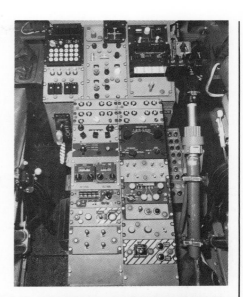

Right: Every square inch of cockpit space in modern aircraft seems to acquire instrumentation, such as the console between the pilots of a Sea King, where most of the radio gear is fitted.
Centre: The complex rotor head for the Sea King's fully articulated five-blade main rotor.
Below: Medevac interior for a Sea King, with six litters and limited seating.

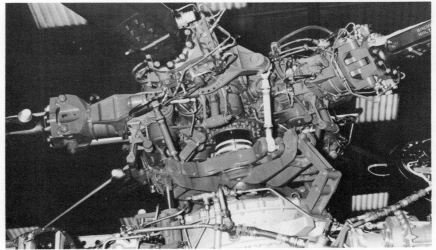

Opposite, top: A Sea King cruises alongside HMS *Ark Royal* before sideslipping in for a landing.

Opposite, below: In the transport role the Sea King has useful capacity, the provision of folding seats permitting the alternative use of the type for freighting without the need for the time-consuming business of seat removal.

Right: The Royal Navy's first Sea King HAS. Mk 2 shows off its lines. This version was basically similar to the Sea King HAS. Mk 1 apart from an uprated powerplant and detail improvements. Surviving Sea King HAS. Mk 2s were later upgraded to Sea King HAS. Mk 5 standard.

Below: Provision of external sling capability allows this Sea King HAS. Mk 2 to move as a single load a 105-mm howitzer and its tow vehicle.

British procurement got under way with four aircraft bought from Sikorsky, one a complete airframe and the other three as kits of components for assembly by Westland. The first delivery was an **S-61 D-2** supplied and flown in civil and military markings (G-ATYU and XV370) with General Electric turboshafts. This aircraft arrived in the UK during October 1966, and was soon followed by the three other aircraft in kit form. These were assembled with Rolls-Royce Gnome H.1400 turboshafts (licence-built versions of the General Electric T58), so few modifications to the engine assembly were required. But whereas the purely American SH-3Ds had used a Hamilton Standard electro-mechanical engine control system, on British-built aircraft a Hawker Siddeley Dynamics full-authority elec-

tronic control system was provided. The three initial Westland-assembled Sea Kings were serialled XV371, XV372 and XV373, and the first of these flew on 8 September 1967. With three more aircraft available for the trials programme, development of the British Sea King series was speeded up considerably in late 1967, the three British-engined aircraft being used for the proving of the Newmark Mk 31 automatic flight control system, for handling and performance trials, and for engine trials. The last mentioned was XV373, and this helicopter was effectively lost on 15 January 1969 after an extremely heavy landing following an auto-rotative descent made necessary by the flaming-out of the Gnome engines as a result of ice ingestion during a poor-weather cross-country flight. (Considerable icing problems had

Multi-mark Sea Kings: starting from the back, these are the Sea King HAS. Mk 5, the Sea King HC. Mk 4, the Sea King HAS. Mk 3, the Sea King HAS. Mk 1 and the Sea King Mk 4X. The last is one of two research helicopters for the RAE Farnborough.

also been encountered with American-built S-61 helicopters, and many of the series were fitted with ice-deflection shields in front of the air inlets.)

By this time production of the initial British version was well under way to meet the balance of the original order for 60 aircraft. This was the **Sea King HAS. Mk 1**, which was by now a somewhat different aircraft in comparison with its American forebears as a result of differences between British and American thinking in tactical ASW matters. As noted above, the US Navy was generally content to use its ASW helicopters as remote sensor platforms for surface warships, which controlled the operation in their Combat Information Centers and then moved in for the actual attack on the submarine. The British, however, had since 1954 been moving towards a tactical philosophy which placed the whole of the ASW mission on the shoulders of the helicopter crew. This philosophy had reached an initial plateau in the Wessex HAS. Mk 3, the Sea King's immediate predecessor and itself a radical turbine-engined development of the S-58 series. In the Wessex HAS. Mk 3, the crew complement was four, consisting of two pilots and a tactical crew of two in the cabin. So far as this went, it was the same as in the SH-3 series. But whereas the two cabin crew in the SH-3 series comprised a pair of sonar operators, in the Wessex HAS. Mk 3 they comprised a sonar operator and a tactical co-ordinator, the cabin being completed as a small operations centre to permit ASW operations entirely independent of surface vessels.

The same configuration was adopted for the Sea King HAS. Mk 1, with the main item in the cabin an automatic tactical plotting display. This shows a fixed search area, always north-orientated by means of an input from the compass, and on it the tactical co-ordinator can call up selective displays using inputs from the sonar and radar systems. The former is a dunking system based on the Plessey Type 195 equipment while the latter, an innovation to the Sea King series in terms of air and surface search capability, is the Ekco AW391 equipment with its antenna in a dorsal radome aft of the power-plant assembly. The final input comes from the Marconi AD580 Doppler navigation system, to ensure that the automatic tactical plotting display is kept fully updated as to the helicopter's exact position, allowing an extremely accurate assessment of

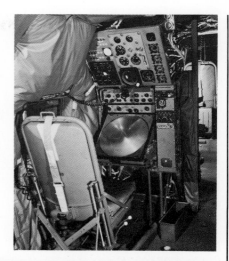

Right: Most important in the Sea King HAR. Mk 3 is the position of radar operator, who is faced with the problems of finding vessels in distress under the most adverse of conditions. Extraordinary rescues during violent storms have won the RAF's SAR service a very high reputation.
Below: A Sea King HAS. Mk 2 of No. 820 Squadron, based on board HMS *Hermes*, shadows a 'Foxtrot' class diesel submarine of the Russian navy, detected on the surface while en route to Cuba in January 1980. Note the half-lowered transducer for the dunking sonar.

the bearings and ranges of all air, surface and underwater craft in the search area.

While the development flying of the Sea King type was pressed ahead in two years of concerted effort with the Sikorsky-provided aircraft, production of the British model was moving ahead smoothly, the first production of Sea King HAS. Mk 1 flying on 7 May 1969 and deliveries to the Royal Navy beginning on 11 August 1969 at a ceremony at the RNAS Culdrose. The last of 56 Sea King HAS. Mk 1s was delivered in May 1972. The Royal Navy's first task was to validate the type under service conditions, and for this purpose No. 700S Squadron was commissioned at Culdrose on 20 August 1969. An intensive

flying trial was conducted over the following nine months, six Sea King HAS. Mk 1s putting in some 2,700 hours of flying, including 2,000 hours devoted to reliability testing. This latter proved the Sea King an admirably reliable aircraft thanks to the combination of excellent basic design and first-class systems well-integrated into an effective whole.

At the end of the intensive flying trial period, No. 700S Squadron's Sea Kings were allocated to No. 737 Squadron at Portland, this being the operational conversion unit for Wessex and Sea King crews, with special emphasis on tactics and weapons. The initial Sea King crews moved to No. 737 Squadron from No. 706 Squadron at Culdrose, this being the advanced flying training unit for the Sea King, the squadron's first aircraft having been received in November 1969. The first three operational Sea King squadrons were No. 824, commissioned aboard HMS *Ark Royal* in February 1970; No. 826, commissioned aboard HMS *Eagle* in April 1970; and No. 819, commissioned at Prestwick in February 1971. Early operational experience confirmed that the Sea King HAS. Mk 1 was an exceptional ASW platform, well liked by its crew not only for its operational capability, but also for its reliability and the knowledge that a single-engine water take-off rating of 1,665 hp (1,242 kW) offered excellent survivability.

Written into Westland's licence agreement with Sikorsky was the right to sell its own version of the Sea King in any part of the world other than North America, though on a non-exclusive basis that meant customers were often faced with Agusta, Sikorsky and Westland versions of the same basic aircraft. (The Japanese Constitution prohibits the export of arms, so there was no threat from Mitsubishi, at least in the export of military versions of the Sea King.) In these circumstances, there could be no surer proof of the Westland Sea King's overall superiority than the greater export success enjoyed by the British version of the West's most important rotary-wing ASW platform.

First off the mark for the Westland Sea King (after the Royal Navy) was the Bundesmarine (West German navy), which in May 1969 placed an order for 22 Sea Kings. These were not intended for ASW, however, but rather to replace the current Grumman HU-16 Albatross amphibians in the search-and-rescue role. Designated **Sea King Mk 41**, the West German version is remarkably

The Sea King Mk 41 fleet of the West German navy has an exceptionally high level of equipment.

similar to the Sea King HAS. Mk 1, apart from the removal of the sonar equipment and features of the automatic flight control system relevant to sonar operations. The two-man 'tactical crew' is retained, though in the West German model the tactical co-ordinator is replaced by a search co-ordinator. More cabin volume was clearly desirable in the SAR role, so the aft bulkhead was shifted 5 ft 8 in (1.73 m) to the rear, providing a cabin length of 24 ft 11 in (7.60 m) for the two-man search crew, 15 seated passengers and three litters, with the remaining volume available for an alternative of six extra seats or three more litters. To provide extra fields of vision for the SAR role, extra windows (hemispherical units) were added on each side of the cabin at the extreme rear, and a winch (identical with the type that can be fitted optionally to Royal Navy Sea Kings) was added over the sliding starboard-side door. Other features of the Sea King Mk 41 are an attachment point for an 8,000-lb (3,629-kg) cargo sling under the fuselage, revised navigation and communications gear, and extra fuel under the rear of the cabin floor to increase maximum capacity to 800 Imp gal (3,637 litres). This means that the Sea King Mk 41 can fly out to a radius of 260 miles (418 km), hover for 30 minutes and winch up 12 survivors and then return to base with untouched reserves. Maximum endurance in the SAR role is 5 hours 45 minutes.

The first Sea King Mk 41 was flown on 6 March 1972, and

soon demonstrated that it met all the West Germans' requirements, allowing production of the rest of the batch to be undertaken. The Sea King Mk 41 helicopters were allocated to Marinefliegergeschwader Nr 1, based at Kiel-Holtenau but with detachments at Borkum, Sylt and Heligoland.

Other customers for the Sea King in the SAR role have been Norway and Belgium, whose air forces operate the **Sea King Mk 43** and **Sea King Mk 48** respectively. Norway ordered 10 Sea King Mk 43s in December 1970, the first of these flying on 19 May 1972. The Sea King Mk 43 is basically similar to the Sea King Mk 41, but has less sophisticated communications gear. The cabin is configured for the accommodation of the two-man SAR crew plus 18 seated passengers or five passengers and six litters, and special heating is provided to ensure a cabin temperature of up to 20°C at external air temperatures of down to —30°C. Also provided is a standard 10-man dinghy, and the Sea King Mk 43 has amply proved its capability to meet the Norwegian requirement of a radius of 230 miles (370 km), with 16 survivors rescued during a 9-minute hover at maximum range. The Sea King Mk 43s are operated by No. 330 Squadron, which is based at Bodo but deploys two-helicopter detachments to Banak, Orland and Sola. When not needed for SAR duties, the Sea King Mk 43s are available for freighting, medevac and other utility roles in the north of the country.

The Belgian order was for five Sea King Mk 48 helicopters, four being configured for SAR duties and the fifth with alternative VIP accommodation. These helicopters are similar to the West German and Norwegian aircraft in basic capability, and all five had been delivered by November 1976, serving from that time with No. 40 Squadron based at Coxyde.

Westland has also exported ASW versions of the Sea King, the lead customers being India and Pakistan, who took **Sea King Mk 42** and **Sea King Mk 45** helicopters respectively. The Indian order, originally for six aircraft (the first flying on 14 October 1970) but later increased by another six, was completed in 1974, the Sea King Mk 42 being very similar to Sea King HAS. Mk 1. These helicopters, as well as **Sea King Mk 42A** and **Sea King Mk 42B** helicopters ordered in later batches, are operated by Nos 330 and 336 Squadrons of the Indian naval air arm from their base at Cochin. The Sea King Mk 42A has a haul-down system permitting the type's use on small warships, and the Sea King Mk 42B is based on the much-improved Sea King HAS. Mk 5 for the Royal Navy. India's total procurement of the Sea King Mk 42 series was 15 helicopters,

The Royal Norwegian air force's Sea King Mk 43 helicopters serve with No. 330 Squadron, operating in detached flights.

of which 14 survived into 1983.

Pakistan ordered six Sea King Mk 45 ASW helicopters, all but identical with the Indian Sea King Mk 42s apart for provision for AM.39 Exocet anti-shipping missiles. All six Sea King Mk 45 helicopters had been delivered by 1975, and these are operated by the Pakistani navy for ASW and anti-shipping duties.

More important, however, was an order for the Royal Australian Navy. To meet this requirement Westland evolved the multi-role and uprated **Sea King Mk 50**, the first example of which made its maiden flight on 30 June 1974. The Royal Australian Navy had already taken 27 examples of the Wessex Mk 31, so Westland's sales effort was centred on a new model of the Sea King with good 'hot and high' performance, and the ability to undertake several roles including ASW, SAR, vertical replenishment of ships at sea, tactical troop movement and medevac, all with the very considerable advantage of self-ferrying capability. Westland's first notion into the solution of the 'hot and high' requirement was the use of two Rolls-Royce Gnome H.1400-2 turboshafts, but it was soon discovered that although the H.1400-2's flat-rated output of 1,600 hp (1,194 kW) would provide the necessary performance, the engine's greater length (resulting from the addition of two extra compressor stages) would entail a considerable redesign of the airframe. The very

acceptable alternative was the Gnome H.1400-1, flat-rated to 1,590 hp (1,186 kW) but of exactly the same dimensions as the standard 1,500-hp (1,119-kW) Gnome H.1400. The gearbox was strengthened to a rating of 2,700 hp (2,014 kW) to cater for the extra power, allowing weight to be increased by 500 lb (227 kg) to 21,000 lb (9,526 kg), and a number of improvements (including a six-blade anti-torque rotor) were incorporated. The dunking sonar chosen by Australia was the AN/AQS-13B. The Royal Australian Navy's first order was for 10 Sea King Mk 50 aircraft, later increased to 12. The ability of this uprated type to carry an internal load of 6,000 lb (2,722 kg) or an external load of 8,000 lb (3,629 kg) proved very useful to the Royal Australian Navy, which started to accept delivery in the autumn of 1974. In 1983 the Sea King Mk 50s were on the strength of No. 817 Squadron (HS-817).

The final export order for the Sea King type was for six aircraft required by the Egyptian navy but bought by Saudia Arabia. These Sea King Mk 47 helicopters are closely akin to the Sea King HAS. Mk 1, and are operated by a squadron of the Egyptian navy from the base at Alexandria.

The experience gained with the Sea King Mk 50 allowed Westland to offer the Royal Navy an improved model of the Sea King, dual-capable ASW and SAR aircraft of which the Royal Navy ordered 21 under the designation **Sea King HAS. Mk 2**. This variant, whose first example flew on 6 September 1977, is based on the Sea King Mk 50. The Royal Navy received 21 new-build Sea King HAS. Mk 2s, and its surviving Sea King HAS. Mk 1s were recycled through the Westland factory at Yeovil for refurbishment and updating to Sea King HAS. Mk 2 standard.

Above: The Egyptian navy's six Sea King Mk 47s are based near Alexandria, and operate in the ASW role with sonar as their main sensor.

Left: Intended mainly for operation from shore bases, the Pakistani navy's Sea King Mk 45s are provided with provision for anti-ship missiles, making them potentially decisive in coastal operations.

Below: The version ordered by the Royal Australian Navy is the Sea King Mk 50, seen here (together with a Wessex) above the now discarded carrier HMAS *Melbourne*. The Sea King Mk 50 is a hybrid type, based on the Sea King HAS. Mk 1 but with a fully uprated powerplant and AQS-13B dunking sonar.

Opposite: The Sea King HAS. Mk 5 for the Royal Navy is an extremely advanced ASW and SAR helicopter with MEL Sea-searcher radar, Decca Type 71 Doppler navigation and LAPADS acoustic processing and display system for signal from sonobuoys.

There followed 16 examples of the **Sea King HAR. Mk 3**, a specialised SAR version for the Royal Air Force's No. 202 Squadron, nominally based at Lossiemouth but normally based as detached pairs at Boulmer, Leconfield, Coltishall and Lossiemouth. The Sea King HAR. Mk 3 combines the uprated dynamic system of the Sea King HAS. Mk 2 with features of the West Germans' Sea King Mk 41, notably the lengthened cabin and bulged rear windows for optimum visual search capability. Cabin accommodation is provided for the air-electronics operator/winch operator and loadmaster/winchman plus six litters or 19 seated survivors.

Westland's latest variant of the Sea King ASW helicopter is the **Sea King HAS. Mk 5**, which offers significantly improved capabilities in comparison with the Sea King HAS. Mk 2. Original plans called for the production of 17 new Sea King HAS. Mk 5 helicopters, and the gradual upgrading of all surviving Sea King HAS. Mk 2 helicopters to this standard, the first few being converted by Westland and the remainder by the Royal Navy itself using Westland-supplied kits. The airframe/powerplant of the Sea King HAS. Mk 5 remains little altered from that of the Sea King HAS. Mk 2, but the electronics and data-processing capability of the type have been very greatly enhanced. The first two Sea King HAS. Mk 5s were delivered in 1980, and early flight trials confirmed that here was the most advanced ASW helicopter in the Western world. Externally, the feature that most distinguishes the Sea King HAS. Mk 5 is the larger dorsal radome for the MEL Sea Searcher radar, which has twice the range of the earlier AW391 equipment and far greater discrimination in conditions of electronic countermeasures. Other features are the Decca 71 Doppler navigation system, the

Marconi LAPADS (Lightweight Acoustic Processing And Display System), and a new suite of data-processing and communications equipment. The LAPADS is a development of a system designed for large fixed-wing aircraft, and the Sea King HAS. Mk 5 can also use information garnered by sonobuoys dropped from BAe Nimrod maritime reconnaissance aircraft. This, coupled with the LAPADS capability, means that the Sea King HAS. Mk 5 can detect, plot and identify submarines with greater speed and assurance of accuracy. The greater volume of this equipment has meant the shifting of the cabin aft bulkhead to the rear by some 6 ft 6 in (1.98 m). Development of the Sea King HAS. Mk 5 began in May 1979, and the availability of the type during the Falklands campaign of 1982 was perhaps decisive in preventing Argentine submarine attacks on the British South Atlantic Task Force; the type's radar was also a strong inducement for the Argentines not to risk surface vessels close to the British force. No Sea Kings were lost to enemy fire, but two Sea King HAS. Mk 5s were lost to other causes during the campaign. So successful was the type in terms of operational capability and reliability that it has subse-

quently been decided to replace the lost helicopters and also to increase the Royal Navy's total purchase of the type by another six units.

The one material shortcoming of the British forces in the South Atlantic that could have spelled total disaster was lack of any effective type of airborne early warning. With a rapidity almost unbelievable in peacetime, a Sea King was rapidly converted as an experimental airborne early warning aircraft with Thorn EMI Searchwater radar. Though the development was too late for service in the South Atlantic, its wide potential use has led to a British decision to procure a small number of such conversions to increase the capability of the British surface forces. The conversion is ingenious and wholly satisfactory: the antenna is housed under an inflatable fabric dome and cantilevered out of the starboard side of the fuselage on a turntable mounting, and the rest of the equipment (signal processing unit, cooling system and operator's console) is situated at the front of the cabin. When the Searchwater equipment is in use, the dome unit is swivelled down to the vertical position, and at all other times the unit is turned to lie horizontally to the rear.

Westland Commando

In 1971 Westland announced that it was starting the development of a transport version of the S-61 design, the object being a utility helicopter optimised for the military transport of men and equipment, but capable of effective use as a logistic transport and medevac aircraft, plus a tertiary level of possible roles in the fields of SAR and ground attack. What emerged was the **Westland Commando**, based closely on the Sea King but without ASW equipment and amphibious capability, the sponsons being removed and a fixed landing gear arrangement substituted. Otherwise the airframe and the powerplant were those of comparable Sea King variants, and the Commando first flew on 12 September 1973. Flight trials confirmed that the Commando could meet its

specification in being able to move 28 troops over a range of 345 miles (555 km) or 8,000 lb (3,629 kg) of freight over a range of 150 miles (241 km). Also revealed was the extreme diversity of the weapons that could be fitted to the type for a variety of tactical tasks associated with ground warfare.

The first order was for five aircraft, designated **Commando Mk 1**, paid for by Saudi Arabia but intended for the Egyptian air force. These were interim aircraft retaining the sponsons of the Sea Kings, and all five were delivered in 1974. Next came the **Commando Mk 2** with definitive landing gear and yet more tactical versatility. The first of this model flew on 16 January 1975, and deliveries included 19 to Egypt, three **Commando Mk 2A** helicopters to Qatar, two

Commando VIP Mk 2B helicopters to Egypt and one **Commando VIP Mk 2C** to Qatar. Also available in the 1980s is the Commando Mk 3, an advanced multi-role armed helicopter.

The UK also bought the Commando type, 15 being ordered in July 1978. Fitted out for the accommodation of a maximum of 27 troops, but more normally operating with 20 fully equipped Royal Marine Commandos or 6,000 lb (2,721 kg) of freight, these helicopters are designated **Sea King HC. Mk 4** in British service. They performed excellently in the Falklands campaign, and though no Sea King HC. Mk 4s were lost to enemy fire, three were lost to other causes. These losses will be made up with new production.

Below: This Commando Mk 2 under test for the Egyptian air force has mass drums outboard of the wheels to simulate the weight of the weapons that would otherwise be carried.

Above: The provision of Searchwater radar is a clumsy but nonetheless effective expedient to provide Royal Navy task forces with airborne early warning of imminent air attacks. Here a Sea King fitted with such radar lands on HMS *Illustrious* in the autumn of 1982.
Below: Operated by the Royal Navy under the designation Sea King HC. Mk 4, the Westland Commando can operate in all climatic conditions from the deserts of the Middle East to the snows of Norway.

The Future

Sikorsky produced some 800 examples of its S-61 series, and licence production by the end of 1982 had added about another 400. These 1,200 helicopters are signally important: in their ASW role the Sea Kings offer the West's navies an unparalleled rotary-wing defence against submarine attack; in their CH-3, HH-3 and Commando forms they are useful and multi-capable transports suitable for a wide variety of roles; in their S-61L and S-61N forms they still play a major part in the small helicopter airline business and the considerably larger offshore-resources exploitation market; and in other forms they play parts in less diverse but nonetheless important civil and military tasks. It seems inevitable, therefore, that the Sea King family will soldier on for some years to come, and that new variants may yet develop, at least from Agusta and Westland.

A Westland Sea King HC. Mk 4 of No. 846 Squadron from HMS *Bulwark*, as indicated by the deck letter 'B' on the fin. All the Royal Navy's Sea King HC. Mk 4 helicopters are operated by Nos 845 and 846 Squadrons, which previously operated the Wessex HU Mk 5 in exactly the same role, namely support for Royal Marine assault forces.

Specifications

Sikorsky SH-3D Sea King technical description

Type: all-weather amphibious anti-submarine helicopter.

Rotor system: five-blade main and anti-torque tail rotors; the blades are interchangeable on each rotor, and each main-rotor blade consists of a D-section aluminium alloy leading edge with honeycomb-filled trailing-edge pockets; the main rotor is fully articulated and oil-lubricated, and each of the five blades is attached to the hub by means of a flanged cuff bolted to a matching flange on the steel rotor head; a rotor brake is standard, and power-folding of the main rotor is provided; the anti-torque rotor is all metal; both rotors are driven through the main gearbox, driven by both engines via steel drive shafts and freewheel units; the main gearbox reduces engine revolutions in the ratio of 93.43:1 for the main rotor, and the provision of intermediate and tail gearboxes for the anti-torque rotor shafting reduces revolutions in the ratio of 16.7:1 for the tail rotor.

Fuselage: all-metal semi-monocoque structure for the amphibious boat hull with single step, fuselage, boom and rotor pylon; single crew door on the port side of the fuselage forward of the sponson, and an aft-sliding door on the starboard side of the fuselage.

Tail unit: all-metal half tailplane on starboard side of the anti-torque rotor pylon.

Landing gear: amphibious capability provided by the single-step boat hull and stabilising sponsons attached to the fuselage sides by horizontal struts and angled shock-absorbing struts; for land operation the helicopter has semi-retractable tailwheel landing gear, comprising a fixed tailwheel of Goodyear manufacture with a 6.00-6 tyre and hydraulically-actuated rearward retracting twin-wheel units (with Goodyear hydraulic wheel brakes and 6.50-10 tyres) attached to the sponsons; pneumatic pop-out emergency flotation bags are attached to the outer sides of the sponsons.

Powerplant and fuel system: two 1,400-shp (1,044-kW) General Electric T58-GE-10 turboshafts are located above the cabin ceiling just forward of the gearbox/main rotor hub assembly, lying side-by-side with their inlets above the port-side door; maximum fuel capacity is 840 US gal (3,180 litres) in three under-floor bladder tanks in the fuselage; the forward tank holds 347 US gal (1,314 litres), the centre tank 140 US gal (530 litres) and the rear tank 353 US gal (1,336 litres), and the refuelling point is on the port side of the fuselage; the oil tank holds a maximum of 7 US gal (26.5 litres).

Accommodation: pilot and co-pilot on the flightdeck with dual controls, and two sonar operators in the cabin.

Electronics and operational equipment: Bendix AN/AQS-13 sonar with dunking transducer; Hamilton Standard automatic stabilisation system with automatic transition into the hover, where the sonar coupler holds the helicopter's altitude automatically with the aid of a radar altimeter and its position automatically with the aid of Teledyne AN/APN-182 Doppler radar; there is provision for a 600-lb (272-kg) rescue winch over the sliding starboard-side door, and for an 8,000-lb (3,629-kg) capacity low-response cargo sling (with automatic touchdown release) for external loads.

Armament: provision for 840 lb (381 kg) of weapons (normally lightweight homing torpedoes or depth charges) under the horizontal sponson struts.

Systems: primary and secondary hydraulic systems with a pressure of 1,500 lb/sq in (103.5 bars) for the flight controls; utility hydraulic system with a pressure of 3,000 lb/sq in (207 bars) for the landing gear, winch and powered blade folding; pneumatic system with a pressure of 3,000 lb/sq in (207 bars) for blow-down emergency landing gear extension; electrical system includes one 300-amp DC generator, two 115-amp 20 kVA AC generators and a 24-volt 22 Ah battery pack; optional auxiliary power unit.

Sikorsky SH-3A Sea King

Type:	anti-submarine helicopter
Accommodation:	two pilots and two systems operators
Armament:	840 lb (381 kg) of torpedoes, depth charges or other stores
Powerplant:	two 1,250-hp (933-kW) General Electric T58-GE-8B turboshafts
Performance:	
maximum speed	153 mph (246 km/h) at sea level
cruising speed	133 mph (214 km/h)
initial climb rate	1,820 ft (555 m) per minute
service ceiling	10,800 ft (3,290 m)
range	625 miles (1,006 km)
Weights:	
empty equipped	11,419 lb (5,180 kg)
normal take-off	18,044 lb (8,185 kg)
maximum take-off	20,500 lb (9,299 kg)
Dimensions:	
span (rotor diameter)	62 ft (18.90 m)
length (fuselage)	54 ft 9 in (16.69 m)
height (overall)	16 ft 10 in (5.13 m)
main rotor disc area	3,019.1 sq ft (280.47 m²)

Sikorsky SH-3D Sea King

Type:	anti-submarine helicopter
Accommodation:	two pilots, and two systems operators or alternative accommodation suiting other roles
Armament:	as for SH-3A
Powerplant:	two 1,400-hp (1,044-kW) General Electric T58-GE-10 turboshafts
Performance:	
maximum speed	166 mph (267 km/h) at sea level
cruising speed	136 mph (219 km/h)
initial climb rate	2,200 ft (670 m) per minute
service ceiling	14,700 ft (4,480 m)
range	625 miles (1,006 km)
Weights:	
empty equipped	12,087 lb (5,483 kg)
normal take-off	18,897 lb (8,572 kg) for ASW
maximum take-off	20,500 lb (9,299 kg)
Dimensions:	
span (rotor diameter)	as for SH-3A
length (fuselage)	
height (overall)	
main rotor disc area	

Sikorsky SH-3H Sea King

Type:	multi-role helicopter
Accommodation:	two pilots and two systems operators (ASW) or two pilots and 15 troops (utility role)
Armament:	as for SH-3A
Powerplant:	as for SH-3D
Performance:	
maximum speed	as for SH-3D
cruising speed	
initial climb rate	
service ceiling	
range	
Weights:	
empty equipped	12,087 lb (5,483 kg)
normal take-off	—
maximum take-off	21,000 lb (9,526 kg)
Dimensions:	
span (rotor diameter)	as for SH-3A
length (fuselage)	
height (overall)	
main rotor disc area	

Sikorsky CH-3E

Type:	amphibious transport helicopter
Accommodation:	two pilots, one cabin crew and 30 troops, or 15 litters
Armament:	none
Powerplant:	two 1,500-hp (1,119-kW) General Electric T58-GE-5 turboshafts

Performance:

maximum speed	162 mph (261 km/h) at sea level
cruising speed	144 mph (232 km/h)
initial climb rate	1,310 ft (399 m) per minute
service ceiling	11,100 ft (3,385 m)
range	465 miles (748 km)

Weights:

empty equipped	13,255 lb (6,012 kg)
normal take-off	21,247 lb (9,638 kg)
maximum take-off	22,050 lb (10,001 kg)
payload	5,000 lb (2,268 kg) of freight

Dimensions:

span (rotor)	62 ft (18.90 m)
length (fuselage)	57 ft 3 in (17.45 m)
height (overall)	18 ft 1 in (5.51 m)
main rotor disc area	3,019.1 sq ft (280.47 m²)

Sikorsky HH-3E
Jolly Green Giant

Type:	armed rescue helicopter
Accommodation:	two pilots and two cabin crew, plus up to 25 troops
Armament:	four 7.62-mm (0.3-in) General Electric Miniguns on flexible mountings
Powerplant:	as for CH-3E

Performance:

maximum speed	164 mph (264 km/h) at sea level
cruising speed	154 mph (248 km/h)
initial climb rate	1,520 ft (463 m) per minute
service ceiling	13,000 ft (3,960 m)
range	480 miles (772 km) on internal fuel or 737 miles (1,186 km) with external fuel

Weights:

empty equipped	13,435 lb (6,094 kg)
normal take-off	19,500 lb (8,845 kg)
maximum take-off	22,050 lb (10,001 kg)
payload	—

Dimensions:

span (rotor)	62 ft (18.90 m)
length (fuselage)	57 ft 3 in (17.45 m) excluding probe
height (overall)	18 ft 1 in (5.51 m)
main rotor disc area	3,019.1 sq ft (280.47 m²)

Sikorsky S-61A

Type:	amphibious transport helicopter
Accommodation:	two pilots and two cabin crew, plus 26 troops, or 15 litters, or 12 VIP passengers
Armament:	none
Powerplant:	as for SH-3A, or two Rolls-Royce Gnome H.1200 turboshafts

Performance:

maximum speed	as for SH-3A
cruising speed	
initial climb rate	
service ceiling	
range	

Weights:

empty equipped	9,763 lb (4,428 kg)
normal take-off	20,500 lb (9,299 kg)
maximum take-off	21,500 lb (9,752 kg)

Dimensions:

span (rotor)	as for SH-3A
length (fuselage)	
height (overall)	
main rotor disc area	

Sikorsky S-61L Mk II

Type:	transport helicopter
Accommodation:	two pilots, one cabin crew and up to 30 passengers
Armament:	none
Powerplant:	two 1,500-hp (1,119-kW) General Electric CT58-140-2 turboshafts

Performance:

maximum speed	146 mph (245 km/h) at sea level
cruising speed	138 mph (222 km/h)
initial climb rate	1,300 ft (396 m) per minute
service ceiling	12,500 ft (3,810 m)
range	265 miles (426 km) with reserves

Weights:

empty equipped	11,701 lb (5,308 kg)
normal take-off	19,000 lb (8,618 kg)
maximum take-off	20,500 lb (9,299 kg)
payload	—

Dimensions:

span (rotor diameter)	62 ft (18.90 m)
length	72 ft 10½ in (22.21 m) with rotors turning
height (overall)	17 ft (5.18 m)
main rotor disc area	3,019.1 sq ft (280.47 m²)

Sikorsky S-61N Mk II

Type:	amphibious transport helicopter
Accommodation:	as for S-61L
Armament:	none
Powerplant:	as for S-61L

Performance:

maximum speed	146 mph (245 km/h) at sea level
cruising speed	138 mph (222 km/h)
initial climb rate	1,300 ft (396 m) per minute
service ceiling	12,500 ft (3,810 m)
range	518 miles (833 km) with reserves

Weights:

empty equipped	12,510 lb (5,674 kg)
normal take-off	19,000 lb (8,618 kg)
maximum take-off	20,500 lb (9,299 kg), or 22,000 lb (9,979 kg) with external load
payload	—

Dimensions:

span (rotor diameter)	62 ft (18.90 m)
length	72 ft 10 in (22.20 m) with rotors turning
height (overall)	18 ft 5½ in (5.63 m)
main rotor disc area	3,019.1 sq ft (280.47 m²)

Sikorsky Payloader

Type:	medium-lift utility helicopter
Accommodation:	two pilots
Armament:	none
Powerplant:	as for S-61L

Performance:

maximum speed	
cruising speed	
initial climb rate	
service ceiling	
range	

Weights:

empty equipped	10,510 lb (4,767 kg)
normal take-off	
maximum take-off	22,000 lb (9,979 kg)
payload	11,000 lb (4,990 kg)

Dimensions:

span (rotor diameter)	as for S-61N
length	
height (overall)	
main rotor disc area	

Agusta-Sikorsky S-61A-4

Type:	utility transport helicopter
Accommodation:	two pilots and one cabin crew, plus up to 31 troops, or 15 litters, or 10 VIP passengers
Armament:	none
Powerplant:	two 1,500-hp (1,119-kW) General Electric CT58-GE-100 turboshafts
Performance:	
maximum speed	generally similar to SH-3D
cruising speed	
initial climb rate	
service ceiling	
range	
Weights:	
empty equipped	generally similar to SH-3D
normal take-off	
maximum take-off	
slung load	
Dimensions:	
span (rotor diameter)	as for SH-3D
length (fuselage)	
height (overall)	
main rotor disc area	

Westland Sea King HAS. Mk 2

Type:	anti-submarine and general-purpose helicopter
Accommodation:	two pilots and two systems operators (ASW) or two pilots and two cabin crew, plus 22 passengers or 9 litters and attendants (general purpose)
Armament:	four Mk 44 torpedoes or four Mk 11 depth charges, plus (general purpose) one 7.62-mm (0.3-in) machine-gun
Powerplant:	two 1,660-hp (1,238-kW) Rolls-Royce Gnome H.1400-1 turboshafts
Performance:	
maximum speed	—
cruising speed	129 mph (208 km/h) at sea level
initial climb rate	2,020 ft (616 m) per minute
service ceiling	4,000 ft (1,220 m) on one engine
range	764 miles (1,230 km) on standard fuel and 937 miles (1,507 km) with auxiliary fuel
Weights:	
empty equipped	13,672 lb (6,201 kg) for ASW
normal take-off	—
maximum take-off	21,000 lb (9,526 kg)
slung load	—
Dimensions:	
span (rotor diameter)	62 ft (18.90 m)
length (fuselage)	55 ft 9¾ in (17.01 m)
height (overall)	16 ft 10 in (5.13 m)
main rotor disc area	3,019.1 sq ft (280.47 m²)

Westland Commando Mk 2

Type:	tactical transport helicopter
Accommodation:	two pilots and up to 28 troops
Armament:	provision for a wide variety of guns, rockets and missiles
Powerplant:	as for Sea King HAS. Mk 2
Performance:	
maximum speed	137 mph (220 km/h) at sea level
cruising speed	129 mph (208 km/h) at sea level
initial climb rate	2,020 ft (616 m) per minute
service ceiling	4,000 ft (1,220 m) on one engine
range	298 miles (480 km) with 28 troops or 120 miles (193 km) with 6,000-lb (2,722-kg) slung load
Weights:	
empty equipped	12,566 lb (5,700 kg)
normal take-off	—
maximum take-off	21,000 lb (9,526 kg)
slung load	6,500 lb (2,948 kg)
Dimensions:	
span (rotor diameter)	62 ft (18.90 m)
length (fuselage)	55 ft 10 in (17.02 m)
height (overall)	16 ft 10 in (5.13 m)
main rotor disc area	3,019.1 sq ft (280.47 m²)

Westland Sea King HAR. Mk 3

Type:	SAR helicopter
Accommodation:	two pilots and two cabin crew (air-electronics/winch operator and loadmaster/winchman) plus survivors
Armament:	none
Powerplant:	as for Sea King HAS. Mk 2
Performance:	
maximum speed	170 mph (274 km/h) at sea level
cruising speed	129 mph (208 km/h)
initial climb rate	390 ft (119 m) per minute vertically
service ceiling	4,000 ft (1,220 m) on one engine
range	764 miles (1,230 km)
Weights:	
empty equipped	12,376 lb (5,613 kg)
normal take-off	—
maximum take-off	21,000 lb (9,526 kg)
Dimensions:	
span (rotor diameter)	as for Sea King HAS. Mk 2
length (fuselage)	
height (overall)	
main rotor disc area	

Sikorsky HH-3F Pelican

Type:	SAR helicopter
Accommodation:	two pilots and two cabin crew
Armament:	none
Powerplant:	as for CH-3E
Performance:	
maximum speed	156 mph (251 km/h) at sea level
cruising speed	150 mph (241 km/h)
initial climb rate	1,240 ft (378 m) per minute
service ceiling	10,500 ft (3,200 m)
range	328-mile (528-km) radius with 20-minute hover and 10 per cent reserves
Weights:	
empty equipped	13,693 lb (6,211 kg)
normal take-off	—
maximum take-off	22,050 lb (10,002 kg)
Dimensions:	
span (rotor diameter)	62 ft (18.90 m)
length (fuselage)	62 ft 6 in (19.05 m) with tail rotor turning
height (overall)	18 ft 1 in (5.51 m)
main rotor disc area	3,091.1 sq ft (280.47 m²)

Acknowledgments

We would particularly like to thank R. G. Carroll and J. P. Belmont of Sikorsky Aircraft in the USA and J. Baird of Westland PLC in the UK for their invaluable help with the pictures for this publication.

Picture research was through Military Archive & Research Services, Braceborough, Lincolnshire and unless otherwise indicated below all material was supplied by Sikorsky and Westland.

Agusta: pp. 36–37.
Bundesmarine: p. 46.
Crown Copyright (FPU): pp. 18 (top right), 18 (bottom), 19 (top), 24 (top).
Crown Copyright (MOD-RN): pp. 17 (inset), 42 (top), 45 (bottom).
US Air Force: p. 26.
US Navy: pp. 12–13.